T0172285

A HANDBOOK OF

Statistical
Analyses
using
S-Plus

Second Edition

A HANDBOOK OF

Statistical Analyses

using

S-Plus

Second Edition

Brian S. Everitt

CHAPMAN & HALL/CRC

A CRC Press Company
Boca Raton   London   New York   Washington, D.C.

Library of Congress Cataloging-in-Publication Data

Everitt. Brian.
  A handbook of statistical analyses using S-PLUS / by Brian S. Everitt.-- 2nd ed.
    p. cm.
  Includes bibliographical references and index.
  ISBN 1-58488-280-8 (alk. paper)
  1. Mathematical statistics--Data processing. 2. S-Plus. I. Title.

QA276.4 .E88 2001
519.5′0285--dc21   2001043682

This book contains information obtained from authentic and highly regarded sources. Reprinted material is quoted with permission, and sources are indicated. A wide variety of references are listed. Reasonable efforts have been made to publish reliable data and information, but the author and the publisher cannot assume responsibility for the validity of all materials or for the consequences of their use.

Neither this book nor any part may be reproduced or transmitted in any form or by any means, electronic or mechanical, including photocopying, microfilming, and recording, or by any information storage or retrieval system, without prior permission in writing from the publisher.

The consent of CRC Press LLC does not extend to copying for general distribution, for promotion, for creating new works, or for resale. Specific permission must be obtained in writing from CRC Press LLC for such copying.

Direct all inquiries to CRC Press LLC, 2000 N.W. Corporate Blvd., Boca Raton, Florida 33431.

Trademark Notice: Product or corporate names may be trademarks or registered trademarks, and are used only for identification and explanation, without intent to infringe.

Visit the CRC Press Web site at www.crcpress.com

© 2002 by Chapman & Hall/CRC

No claim to original U.S. Government works
International Standard Book Number 1-58488-280-8
Library of Congress Card Number 2001043682
Printed in the United States of America   2 3 4 5 6 7 8 9 0
Printed on acid-free paper

Contents

Preface

Since the first edition of this handbook was published in 1994 the development of S-PLUS has continued apace, and a flexible and convenient "point-and-click" facility has now been added to supplement the very powerful command language. In addition, many new methods of analysis and new graphical procedures have been implemented. The changes made in this second edition reflect these changes in the software. Most chapters have been completely rewritten and many new examples are included. And, some of the more embarassing code from the first edition are now excluded. A mixture of the S-PLUS command language and the S-PLUS Graphical User Interface (GUI) is used throughout the book so that readers can become familiar with using both. An appendix gives a relatively concise account of the command language.

It is hoped that this new edition will prove useful to applied statisticians, statistics students, and researchers in many disciplines who wish to learn about the many exciting possibilities for dealing with data presented by the latest versions of S-PLUS, S-PLUS 2000, and S-PLUS 6. All the data sets used in the text are available in the form of S-PLUS data frames from:

www.iop.kcl.ac.uk/IoP/Departments/BioComp/SPLUS.stm

Script files giving the command language used in each chapter are also available from the same address. (Comments given in the text versions are not included in these files.)

Thanks are due to Ms. Harriet Meteyard for her typing of the manuscript and general support during the writing of this book.

B.S. Everitt
June 2001

Distributors for S-PLUS

In the United Kingdom, S-PLUS is distributed by

Insightful
Knightway House
Park Street
Bagshot, Surrey
GU19 5AQ
United Kingdom
Tel: +44 (0) 1276 450 111
Fax: +44 (0) 1276 451 224
sales@uk.insightful.com

In the United States, the distributors are

Insightful Corporation
1700 Westlake Avenue North
Suite 500
Seattle, WA 98109-3044
USA
Tel: (206) 283-8802
Fax: (206) 283-8691
info@insightful.com
Web address: www.insightful.com

Dedication

To my daughters, Joanna and Rachel and
my grandsons, Hywel and Dafydd

Dedication

Chapter 1

An Introduction to S-PLUS

1.1 Introduction

S-PLUS is a language designed for data analysis and graphics developed at AT&T's Bell Laboratories. It is described in detail in Becker et al. (1988), Chambers and Hastie (1993), Venables and Ripley (1997), and Krause and Olson (2000). In addition to providing a powerful language, the most recent versions of the software, S-PLUS 2000 and S-PLUS 6, also include an extensive graphical user interface (GUI) on Windows platforms (this is not available in UNIX). The GUI allows routine (and some not so routine) analyses to be carried out by simply completing various "dialog boxes," and graphs to be produced and edited by a "point-and-click" approach.

In this chapter we introduce both the GUI and the command line language, although details of the former will be left for the remaining chapters of the book, and of the latter, for Appendix A.

1.2 Running S-PLUS

On a Windows platform S-PLUS is opened by double-clicking into the file (or shortcut for) S-PLUS.exe. The result is an S-PLUS window containing a Commands window and/or an Object Explorer window. During an S-PLUS session, Graphics windows may be opened and often output will

1

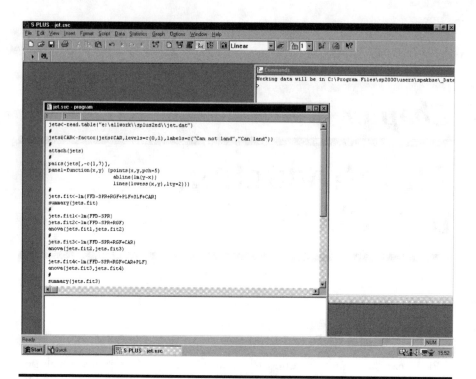

Figure 1.1   Windows seen during a typical S-PLUS session: Command window and script file are shown.

be sent to a Report window, this being opened by changing the default for text output routing in the Options list. The windows seen in a typical S-PLUS session are shown in Figure 1.1. At the top, below the S-PLUS title bar, is the menu bar. On the line below that is the tool bar.

S-PLUS provides a language for the manipulation of 'objects' such as vectors and matrices; commands can be typed into the Commands window next to the > prompt, and any resulting output will appear below, also in the Commands window, unless the Report window option has been selected. If a single command extends over one line of input, the > prompt changes to the plus sign, +. The contents of S-PLUS objects may be viewed by simply typing the name of the object.

The Object Explorer window displays objects of the current session by object category. This window can be opened by clicking into

At the end of a session, the user can select which objects created within the session should be saved within the 'current directory' or database. By default, this is the _data subdirectory of the directory where the S-PLUS files are located, for example, in C:\Program Files\sp6_data. The command **search** lists the current directory under [1].

Since it is usually preferable to keep the data for different projects in different directories, it is a good idea to start an S-PLUS session by setting the directory in which any objects are to be saved and which may contain relevant objects from a previous session. This is done by 'attaching' the directory at the first position of the search path using the command:

```
>attach("c:/project/_data",pos=1)
```

Note that forward slashes are used in the directory path rather than the usual backward slashes. Alternatively, two backward slashes may also be used.

1.3   The S-PLUS GUI: An Introduction

Use of the GUI involves menus, dialog boxes, and point-and-click graphics. For example, many statistical techniques can be applied in S-PLUS by using the **Statistics Menu** and then filling in the relevant dialog box. These boxes have many features in common as we shall see throughout the text. As an example we can look at the **Linear Regression** dialog. This is made available as follows;

- Click on **Statistics**.
- Select **Regression**.
- Select **Linear**.

The resulting dialog is shown in Figure 1.2.

To use the box to carry out a regression analysis would involve filling in the various sections of the box and requesting various options under the **Results**, **Plot**, or **Predict** tabs, as we shall illustrate in detail in Chapter 4.

The GUI approach to producing S-PLUS graphics is extensive and flexible, and can involve either the use of dialog boxes from the **Graphics Menu**, or the **Graphics palettes**. For example, to access the **Scatter Plot** dialog, click on **Graph** in the tools bar, select **2D** and **Scatter Plot** is highlighted by default. Click **OK** and the dialog box shown in Figure 1.3 appears.

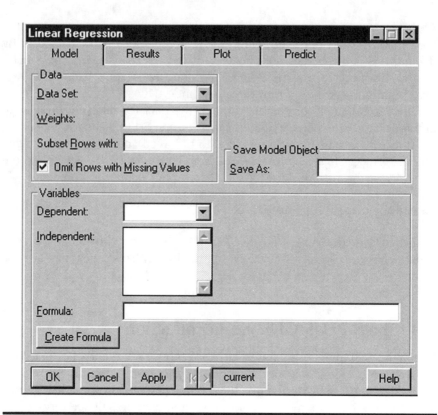

Figure 1.2   Linear Regression dialog.

Again, to use the box to produce a scatterplot would involve filling out the box appropriately; examples will be given in later chapters.

The 2D and 3D palettes are accessed by clicking on

or

respectively, and are shown in Figure 1.4. These can be used after selecting the required data set, to give a wide variety of graphics as we shall illustrate later.

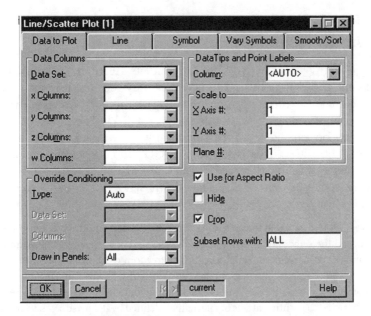

Figure 1.3   Line/Scatter Plot dialog.

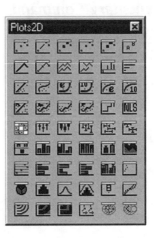

Figure 1.4   2D and 3D graphical palettes.

1.4 The S-PLUS Command Language: An Introduction

Although many users of S-PLUS will find its relatively recent GUI both convenient and sufficient for the analyses they require, it is the command language that remains the main reason that the software is so powerful and flexible. Familiarity with this aspect of S-PLUS enables customised analyses to be carried out relatively simply as we shall illustrate in later chapters. The differences between the GUI and the command language approaches and the advantages of the latter are nicely summarized in the following remarks made by a statistician who has been involved with computers for over 40 years, John Nelder:

> I am very much aware that for the modern student the menu mode is the one preferred, and indeed the only one known. I am, however, not convinced that the menu mode is optimum for all users or for all usages. The freedom of being able to say what you want, instead of responding to given lists, is to me worth having. Imagine how restrictive conversation would be, if instead of making your own points for yourself, you were restricted to pointing at sets of alternatives defined by the person you were talking to. The frustrations would soon become apparent.

In this section we shall simply introduce a few of the most important features of the S-PLUS command language, leaving a more detailed account to be given in Appendix A.

1.4.1 Elementary Commands

Elementary commands consist of either expressions or assignments. For example, typing the expression

```
>42+8
```

in the commands window and pressing return will produce the following output

```
[1] 50
```

Instead of evaluating just an expression, we can assign the value to a scalar, for example,

```
>x<-42+8
>x
[1] 50
```

1.4.2 Vectors and Matrices

Vectors may be created in several ways, the most common is via the concatenate function, **c**, which combines all values given as arguments to the function as a vector

```
>x<-c(1,2,3,4)
>x
[1] 1 2 3 4
```

(Note that S-PLUS is case sensitive, *x* and **X**, for example, are regarded as different objects.)

Arithmetic operations between two vectors return a vector whose elements are the results of applying the operation to the corresponding elements of the original vectors. We can also apply mathematical functions to vectors; the functions are simply applied to each element of the vector

```
>x<-c(1,2,3)
>y<-c(4,9,16)
>x*y
[1] 4 18 48
>sqrt(y)
[1] 2 3 4
```

Matrix objects are frequently needed in S-PLUS and can be created by use of the **matrix** function. For details of this function see Appendix A or use **help(matrix)** in S-PLUS. (Similar help files are available for all S-PLUS functions.)

```
>x<-c(1,2,3)
>y<-c(4,5,6)
>xy<-matrix(c(x,y),nrow=2)
>xy
```

	[,1]	[,2]	[,3]
[1,]	1	3	5
[2,]	2	4	6

1.4.3 Subsetting Matrices and Vectors

S-PLUS has two logical values, **T** (true) and **F** (false), and a number of logical operations that are extremely useful in choosing particular elements from vectors and matrices. (The logical operations are listed in Appendix A.) We can use a logical operator to assign logical values:

```
>x<-3==4
>x
[1] F

>x<-3<4
>x
[1] T

>x<-c(1,2,3,4,5)
>x<4
[1] T T T F F
```

A logical vector can be used to extract a subset of elements from another vector as follows:

```
>x[x<4]
[1] 1 2 3
```

Here the elements of the vector less than 4 are selected as the values corresponding to T in the vector *x*.

We can also select elements in *x* depending on the values in another vector *y*

```
>x<-c(1,2,3,4,5,6,7,8,9,10)
>y<-c(0,0,6,4,3,1,0,0,1,0)
>x[y==0]

[1] 1 2 7 8 10
```

1.4.4 Other S-PLUS Objects

A number of other important S-PLUS objects are mentioned briefly here and in detail in Appendix A. First **list** objects that allow other S-PLUS objects to be linked together, for example,

```
>x<-c(1,2,3)
>y<-matrix(c(1,2,3,4),nrow=2)
>xylist<-list(x,y)
>xylist
```

```
[[1]]
[1] 1 2 3

[[2]]
      [,1]   [,2]
[1,]    1      2
[2,]    3      4

>xylist$x
[1] 1 2 3
```

Note the two alternatives for referring to elements in a list; either the 'double bracket' nomenclature or the **$name** nomenclature can be used.

Secondly, data frames that allow numerical and character vectors to be bound together are the most useful way of storing sets of data. Creating a data frame is described in detail in Appendix A, but as a simple example:

```
>height<-c(50,60,70)
>weight<-c(100,120,140)
>age<-c(20,40,60)
>names<-c("Bob","Ted","Alice")
>data<-data.frame(names,age,height,weight)
>data
```

	names	age	height	weight
1	Bob	20	50	100
2	Ted	40	60	120
3	Alice	60	70	140

A data frame can be used in S-PLUS by first 'attaching' it, using

```
>attach(data)
```

In this way, variables in the data frame can now be conveniently referred to by name.

```
>age
[1] 20 40 60
```

The S-PLUS language also provides the facility for creating functions for specific analyses of interest. Details are given in Appendix A and examples will be given in subsequent chapters.

Although commands can be typed into the commands window, it is far more convenient to use a **script file** (*.ssc), which is an ASCII text file that may be opened within S-PLUS to build up and keep a sequence of commands being used to analyse a particular data set, or indeed several data sets. In this way an entire analysis can be repeated at the press of a button if necessary, for example, if a data entry error is detected. The whole script file may be executed by selecting **Script** and **Run** from the menu bar or by pressing **F10**. Alternatively, one or more commands may be selected and run by highlighting the relevant text within the Script file and pressing the triangle insert button,

Script files can be commented by using the hash symbol, #, at the beginning of a line of text; S-PLUS ignores such lines. (In all but this chapter and Appendix A, we shall assume that commands are being run from a script file and, therefore, will dispense with the > before each command seen when using the commands window.) To open a script file, click on **File** in the menu bar, select **New**, and then highlight **Script File** in the list that appears.

1.5   An Example of an S-PLUS Session

As with any software, the easiest way to learn about S-PLUS is to use it, and this section attempts to give readers a preview of how S-PLUS is used in practice, which they can follow before reaching the more demanding material in subsequent chapters. Here we shall use both the GUI and the command language approaches to carry out some relatively straightforward analyses of the data shown in Table 1.1, which were originally given in Stanley and Miller (1979) and are also reproduced in Hand et al. (1994). (A more-detailed analysis of these data will be made in Chapter 4.)

We shall assume that the data in Table 1.1 are already available as an S-PLUS data frame object, **jets**. Details of data frames and how they are created from the raw data are given in Appendix A. By typing **jets** in the command window and hitting return its contents will be displayed — see Table 1.2. Initially it is sensible to attach the data frame using

```
>attach(jets)
```

To begin learning about the data we might want to look at some suitable summary statistics for each variable; for this we can use the S-PLUS **summary** function.

```
>summary(jets)
```

Table 1.1   Data on Jet Fighters

	Type	FFD	SPR	RGF	PLF	SLF	CAR
1	FH-1	82	1.468	3.30	0.166	0.10	0
2	FJ-1	89	1.605	3.64	0.154	0.10	0
3	F-86A	101	2.168	4.87	0.177	2.90	1
4	F9F-2	107	2.054	4.72	0.275	1.10	0
5	F-94A	115	2.467	4.11	0.298	1.00	1
6	F3D-1	122	1.294	3.75	0.150	0.90	0
7	F-89A	127	2.183	3.97	0.000	2.40	1
8	XF10F-1	137	2.426	4.65	0.117	1.80	0
9	F9F-6	147	2.607	3.84	0.155	2.30	0
10	F-100A	166	4.567	4.92	0.138	3.20	1
11	F4D-1	174	4.588	3.82	0.249	3.50	0
12	F1F-1	175	3.618	4.32	0.143	2.80	0
13	F-101A	177	5.855	4.53	0.172	2.50	1
14	F3H-2	184	2.898	4.48	0.178	3.00	0
15	F-102A	187	3.880	5.39	0.101	3.00	1
16	F-8A	189	0.455	4.99	0.008	2.64	0
17	F-104B	194	8.088	4.50	0.251	2.70	1
18	F-105B	197	6.502	5.20	0.366	2.90	1
19	YF-107A	201	6.081	5.65	0.106	2.90	1
20	F-106A	204	7.105	5.40	0.089	3.20	1
21	F-4B	255	8.548	4.20	0.222	2.90	0
22	F-111A	328	6.321	6.45	0.187	2.00	1

FFD first flight date, in month after January 1940
SPR specific power, proportional to power per unit weight
RGF flight range factor
PLF payload as a fraction of gross weight of aircraft
SLF sustained load factor
CAR a binary variable that takes the value 1 if the aircraft
can land on a carrier, and 0 otherwise.

The output resulting from these commands is shown in Table 1.3. (Like many S-PLUS functions, **summary** is generic, meaning that it can be used to process many different classes of data and give results appropriate to each particular class. Further examples will be given in subsequent chapters.)

Summary statistics for the data in **jets** can also be found by using the GUI as follows:

- Click **Statistics**.
- Select **Data Summaries**.
- Select **Summary Statistics**.

Table 1.2 The Jets Data Frame

> jets

	Type	FFD	SPR	RGF	PLF	SLF	CAR
1	FH-1	82	1.468	3.30	0.166	0.10	Cannot land
2	FJ-1	89	1.605	3.64	0.154	0.10	Cannot land
3	F-86A	101	2.168	4.87	0.177	2.90	Can land
4	F9F-2	107	2.054	4.72	0.275	1.10	Cannot land
5	F-94A	115	2.467	4.11	0.298	1.00	Can land
6	F3D-1	122	1.294	3.75	0.150	0.90	Cannot land
7	F-89A	127	2.183	3.97	0.000	2.40	Can land
8	XF10F-1	137	2.426	4.65	0.117	1.80	Cannot land
9	F9F-6	147	2.607	3.84	0.155	2.30	Cannot land
10	F-100A	166	4.567	4.92	0.138	3.20	Can land
11	F4D-1	174	4.588	3.82	0.249	3.50	Cannot land
12	F1F-1	175	3.618	4.32	0.143	2.80	Cannot land
13	F-101A	177	5.855	4.53	0.172	2.50	Can land
14	F3H-2	184	2.898	4.48	0.178	3.00	Cannot land
15	F-102A	187	3.880	5.39	0.101	3.00	Can land
16	F-8A	189	0.455	4.99	0.008	2.64	Cannot land
17	F-104B	194	8.088	4.50	0.251	2.70	Can land
18	F-105B	197	6.502	5.20	0.366	2.90	Can land
19	YF-107A	201	6.081	5.65	0.106	2.90	Can land
20	F-106A	204	7.105	5.40	0.089	3.20	Can land
21	F-4B	255	8.548	4.20	0.222	2.90	Cannot land
22	F-111A	328	6.321	6.45	0.187	2.00	Can land

The dialog box shown in Figure 1.5 appears. In the **Data Set** window choose **jets,** highlight all but **Type** in the **Variables** window and click **OK**; the results shown in Table 1.4 appear in a Report file which might be printed or copied and pasted into another application.

Perhaps separate summary statistics are required for the class of fighters that can land on a carrier and those that cannot. If so, they can be obtained by highlighting all but CAR in the **Variables** section of the **Summary Statistics** dialog and then highlighting **CAR** and **Type** in the **Group Variables** section. This leads to the results shown in Table 1.5. (Other summary statistics, for example, measures of *skewness* and *kurtosis*, can be requested simply by clicking on the **Statistics** tab of the **Summary Statistics** dialog.)

A t-test for the difference in the population mean values of, say, the variable FFD for planes that can land and cannot land on a carrier can be calculated using the **Two-sample t-test dialog** which is accessed as follows:

Table 1.3 Summary Statistics for the Jet Fighter Data

Type	*FFD*	*SPR*	*RGF*
YF-107A: 1	Min.: 82.0	Min.:0.455	Min.:3.300
XF10F-1: 1	1st Qu.:123.2	1st Qu.:2.172	1st Qu.:4.005
FJ-1: 1	Median:174.5	Median:3.258	Median:4.515
FH-1: 1	Mean:166.3	Mean:3.944	Mean:4.577
F9F-6: 1	3rd Qu.:192.8	3rd Qu.:6.025	3rd Qu.:4.972
F9F-2: 1	Max.:328.0	Max.:8.548	Max.:6.450
(Other):16			

PLF	*SLF*	*CAR*
Min.:0.0000	Min.:0.100	Cannot land:11
1st Qu.:0.1223	1st Qu.:1.850	Can land:11
Median:0.1605	Median:2.670	
Mean:0.1683	Mean:2.265	
3rd Qu.:0.2132	3rd Qu.:2.900	
Max.:0.3660	Max.:3.500	

- Click on **Statistics**.
- Select **Compare Samples**.
- Select **Two Samples, t test**.

Again select the **jets** data set, highlight **FFD** as Variable 1 and **CAR** as Variable 2, then tick the button that shows Variable 2 as a **Grouping variable**. The results shown in Table 1.6 appear in a Report file. (With such a small sample it may be more appropriate to use the Wilcoxon rank sum test rather than the t-test; we leave this as an exercise for the reader since the steps are essentially identical to those described above.) With the command language, the same results can be found using

```
>t.test(FFD[CAR=="Can Land"],FFD[CAR=="Cannot land"]
```

Graphics are an essential component in the analysis of any data set, and a vast range of graphics are available when using S-PLUS, as we shall see in subsequent chapters. Here, however, we consider only the construction of a simple scatterplot. Using the GUI, we proceed as follows:

- Click on **Graph**.
- Select **2D Plot**.

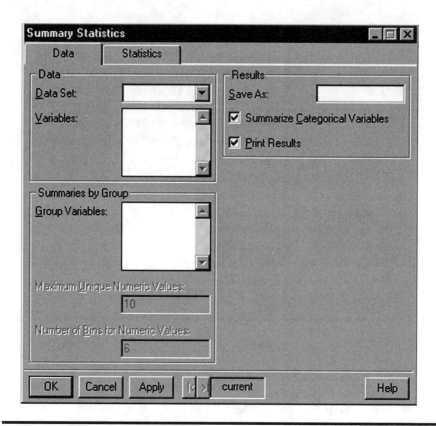

Figure 1.5   Summary Statistics dialog.

The **Insert Graph** menu appears. Since **Scatter Plot** is highlighted by default, simply click **OK** to arrive at the **Line/Scatter Plot** dialog. Select the **jets** data set and, say, **SPR** as the *x* column (the '*x* variable') and **FFD** as the *y* column (the '*y* variable'). Click **OK** to see the scatterplot of these two variables shown here in Figure 1.6. A more interesting scatterplot would be one with the points labelled by type of aircraft. This can again be constructed from the **Line/Scatter Plot dialog.**

We first repeat the steps used to obtain Figure 1.7, but now also select **Type** in the *z* Columns box, and then click the **Symbol** tab of the dialog. Tick **Use Text As Symbol** and in the **Text to Use** box select *z* column. The height of the plotting symbol might also be increased to, say, 0.15. Clicking **OK** now produces the scatterplot shown in Figure 1.7.

Finally, it might be useful to have Figure 1.8 alongside the corresponding scatterplot in which the points are labelled by whether the aircraft can or cannot land on a carrier. This diagram is obtained as follows:

Table 1.4   Summary Statistics for Jet Fighter Data

*** Summary Statistics for data in: jets ***

$$$"Factor Summaries":
      CAR
Cannot land:11
   Can land:11

$$$"Numeric Summaries":

	FFD	SPR	RGF	PLF	SLF
Min:	82.00000	0.455000	3.3000000	0.00000000	0.100000
1st Qu.:	123.25000	2.171750	4.0050000	0.12225000	1.850000
Mean:	166.27273	3.944455	4.5772727	0.16827273	2.265455
Median:	174.50000	3.258000	4.5150000	0.16050000	2.670000
3rd Qu.:	192.75000	6.024500	4.9725000	0.21325000	2.900000
Max:	328.00000	8.548000	6.4500000	0.36600000	3.500000
Total N:	22.00000	22.000000	22.0000000	22.00000000	22.000000
NA's:	0.00000	0.000000	0.0000000	0.00000000	0.000000
Std Dev.:	56.94122	2.367226	0.7529888	0.08665541	1.003312

- With Figure 1.7 constructed and visible, click on **Insert** in the tool bar and select **Graph**.
- Click **OK** on the **Insert Graph** menu that appears.
- Select the **jets** data set and **SPR** as *x* and **FFD** as *y*.
- Here, however, select **CAR** as *z*.
- Click the **Symbol** tab and repeat the appropriate steps described above.
- Click on **OK**.

The resulting diagram is shown in Figure 1.8. (It may have been sensible here to have kept the height of the plotting symbol at its default value for the second diagram; we leave this as an exercise for the reader.)

With the command language the scatterplot in Figure 1.7 is obtained from

```
>plot(SPR,FFD)
```

and the plot in Figure 1.8 from

```
> plot(SPR,FFD,type="n")
> text(SPR,FFD,labels=as.character(Type))
```

Many other examples of the use of the **plot** function will be presented in later chapters.

Table 1.5   Summary Statistics for Jet Fighter Data by Whether or Not Plane Can Land on Carrier

*** Summary Statistics for data in: jets ***

CAR: Cannot land

	FFD	SPR	RGF	PLF	SLF
Min:	82.00000	0.455000	3.3000000	0.00800000	0.100000
1st Qu.:	114.50000	1.536500	3.7850000	0.14650000	1.000000
Mean:	151.00000	2.869182	4.1554545	0.16518182	1.921818
Median:	147.00000	2.426000	4.2000000	0.15500000	2.300000
3rd Qu.:	179.50000	3.258000	4.5650000	0.20000000	2.850000
Max:	255.00000	8.548000	4.9900000	0.27500000	3.500000
Total N:	11.00000	11.000000	11.0000000	11.00000000	11.000000
NA's:	0.00000	0.000000	0.0000000	0.00000000	0.000000
Std Dev.:	51.03724	2.203758	0.5261248	0.07103354	1.200515

CAR: Can land

	FFD	SPR	RGF	PLF	SLF
Min:	101.00000	2.168000	3.9700000	0.0000000	1.000000
1st Qu.:	146.50000	3.173500	4.5150000	0.1035000	2.450000
Mean:	181.54545	5.019727	4.9990909	0.1713636	2.609091
Median:	187.00000	5.855000	4.9200000	0.1720000	2.900000
3rd Qu.:	199.00000	6.411500	5.3950000	0.2190000	2.950000
Max:	328.00000	8.088000	6.4500000	0.3660000	3.200000
Total N:	11.00000	11.000000	11.0000000	11.0000000	11.000000
NA's:	0.00000	0.000000	0.0000000	0.0000000	0.000000
Std Dev.:	60.75255	2.089900	0.7227926	0.1034527	0.642580

Table 1.6   Results of t-Test for Difference in FFD for Planes that Can and Cannot Land on a Carrier

Standard Two-Sample t-Test
data: x: FFD with CAR = Cannot land, and y: FFD with CAR = Can land
t = −1.2768, df = 20, p-value = 0.2163
alternative hypothesis: true difference in means is not equal to 0
95 percent confidence interval:
 −80.44900   19.35809
sample estimates:
 mean of x   mean of y
      151   19.35809

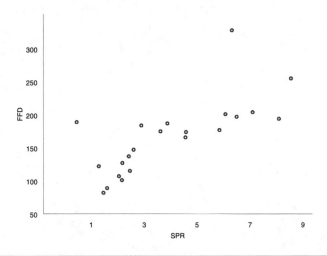

Figure 1.6   Scatterplot of SPR and FFD variables in jets data frame.

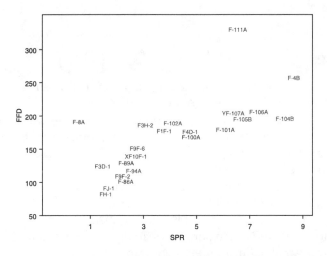

Figure 1.7   Scatterplot of SPR and FFD variables in jets data frame with points labelled by type of aircraft.

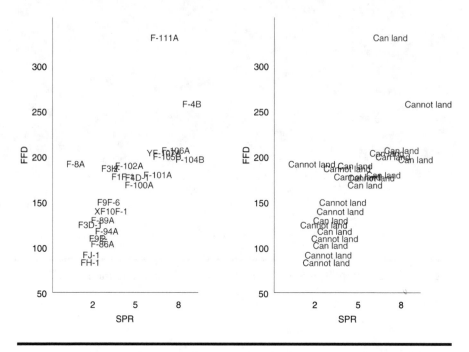

Figure 1.8 Side-by-side scatterplots for the SPR and FFD variables in the jets **data frame.**

Exercises

1.1. Investigate the use of the **apply** function to find the means of the numerical values in the **jets** data frame.

1.2. Use the **rep** function to help produce a new two-column data frame for the jet fighter data in which the numerical variable values for all planes are arranged in a single vector with the type of plane rearranged accordingly.

1.3. Use the **boxplot** function to construct box plots of the values of each numerical variable in the **jets** dataframe.

1.4. Explore the use of the **hist** and **density** functions for plotting histograms and calculating probability density estimates for some of the variables in the **jets** data frame.

1.5. Use the **help** function to find out about the **pairs** function and then apply this function to the variables in the **jets** data frame in ways that you think might be useful.

Chapter 2

Describing Data and Assessing Distributions: Husbands and Wives

2.1 Introduction

The data to be used in this chapter consist of five variables recorded on a random sample of 100 married men and their wives. The five variables are

- **husbage**: husband's age in years
- **husbht**: husband's height in mm
- **wifeage**: wife's age in years
- **wifeht**: wife's height in mm
- **husbagem**: husband's age at the time of the marriage

The data are given in Table 2.1. The label NA is used in S-PLUS to denote a missing value, here generally the result of the wife declining to give her age!

We shall use these data to illustrate some the features of S-PLUS for describing data and for assessing distributions.

Table 2.1   Data on Husbands and Wives

husbage	husbht	wifeage	wifeht	husbagem
40	1659	30	1620	38
58	1616	52	1420	30
32	1695	27	1660	23
42	1753	NA	1635	30
31	1685	23	1610	26
40	1713	39	1610	23
35	1736	32	1700	31
45	1715	NA	1522	41
35	1785	33	1680	24
47	1758	43	1630	24
38	1725	40	1600	31
45	1764	NA	1689	24
50	1674	45	1640	25
27	1700	25	1580	21
28	1721	25	1650	23
37	1829	35	1670	22
56	1710	55	1600	44
27	1745	23	1610	25
47	1809	43	1620	25
31	1585	23	1570	28
35	1705	35	1580	25
27	1721	NA	1560	26
45	1739	39	1610	25
59	1699	52	1440	27
43	1825	52	1570	25
48	1704	NA	1635	27
54	1679	53	1560	NA
43	1755	42	1590	20
54	1713	50	1600	23
61	1723	64	1490	26
51	1585	NA	1504	50
54	1724	53	1640	20
37	1620	39	1650	21
55	1764	45	1620	29
57	1738	55	1560	24
34	1700	32	1640	22
45	1804	41	1670	27
55	1664	43	1760	31
55	1788	51	1600	26
44	1715	41	1570	24

Table 2.1 (Continued) Data on Husbands and Wives

husbage	husbht	wifeage	wifeht	husbagem
42	1731	37	1580	23
34	1760	34	1700	23
45	1559	35	1580	34
48	1705	45	1500	28
44	1723	44	1600	41
59	1700	47	1570	39
64	1660	57	1620	32
34	1681	33	1410	22
37	1803	38	1560	23
49	1884	46	1710	25
63	1705	60	1580	27
48	1780	47	1690	22
64	1801	55	1610	37
33	1795	45	1660	17
52	1669	47	1610	23
27	1708	24	1590	26
33	1691	32	1530	21
46	1825	47	1690	23
27	1949	NA	1693	25
50	1685	NA	1580	21
42	1806	NA	1636	22
54	1905	46	1670	32
49	1739	42	1600	28
62	1736	63	1570	22
34	1845	32	1700	24
53	1736	NA	1555	30
32	1741	NA	1614	22
59	1720	56	1530	24
55	1720	55	1590	21
62	1629	58	1610	23
42	1624	38	1670	22
50	1653	44	1690	35
51	1620	44	1650	30
58	1736	50	1540	32
28	1691	23	1610	23
45	1753	43	1630	21
57	1724	59	1520	24
27	1725	21	1550	24
54	1630	NA	1570	34
25	1815	26	1650	20

Table 2.1 (Continued)   Data on Husbands and Wives

husbage	husbht	wifeage	wifeht	husbagem
57	1575	57	1640	20
61	1749	63	1520	21
25	1705	23	1620	24
32	1875	NA	1744	22
37	1784	NA	1647	22
45	1584	NA	1615	29
44	1790	40	1620	24
52	1798	53	1570	25
60	1725	60	1590	21
36	1685	32	1620	25
35	1664	NA	1539	22
50	1725	49	1670	23
57	1694	55	1620	24
38	1691	38	1530	20
30	1880	31	1630	22
50	1723	47	1650	25
20	1786	18	1590	19
51	1675	45	1550	25
40	1823	39	1630	23
59	1720	56	1530	24

2.2 Some Basic Summaries

The analysis of most data sets begins with the calculation of suitable numerical summary statistics such as variable means and standard deviations, and relatively simple graphics such as *histograms* and *boxplots* describing variable distributions. In the case of multivariate data such as those in Table 2.1, correlations between variables may also be computed and *scatterplots* of pairs of variables constructed.

It may also be necessary to assess whether the individual variables are normally distributed prior to any analysis that makes this assumption. In addition it may be required to check whether the complete set of variables jointly have a multivariate normal distribution. In both cases one approach is to use *probability plots*, as we shall see later in the chapter.

2.3 Analysis Using S-PLUS

We assume that the data in Table 2.1 are available as the S-PLUS data frame **huswif**, with the variables labelled as shown in the previous section. To obtain the basic summary statistics of the data we can again use the **Summary Statistics** dialog as illustrated in Chapter 1; here, however, we will also request measures of skewness and kurtosis. The results are shown in Table 2.2. We see that 16 of the 17 missing values in the data occur for the age of wife variable. Husband's age at marriage has a relatively high degree of skewness.

Table 2.2   Summary Statistics for Husbands and Wives Data

*** Summary Statistics for data in: huswif ***

	husbage	husbht	wifeage	wifeht
Min:	20.0000000	1559.0000000	18.0000000	1410.0000000
1st Qu.:	35.0000000	1691.0000000	33.7500000	1570.0000000
Mean:	44.6400000	1727.5500000	42.3690476	1605.0800000
Median:	45.0000000	1723.0000000	43.0000000	1610.0000000
3rd Qu.:	54.0000000	1764.0000000	52.0000000	1641.7500000
Max:	64.0000000	1949.0000000	64.0000000	1760.0000000
Total N:	100.0000000	100.0000000	100.0000000	100.0000000
NA's:	0.0000000	0.0000000	16.0000000	0.0000000
Std Dev.:	11.1169358	71.8870291	11.4258002	62.8889803
Skewness:	0.1539845	0.2938824	0.1732601	0.4584884
Kurtosis:	−1.0105144	0.6896657	−0.7573466	0.9466165

	husbagem
Min:	17.000000
1st Qu.:	22.000000
Mean:	25.868687
Median:	24.000000
3rd Qu.:	27.500000
Max:	50.000000
Total N:	100.000000
NA's:	1.000000
Std Dev.:	5.674228
Skewness:	1.803211
Kurtosis:	3.862735

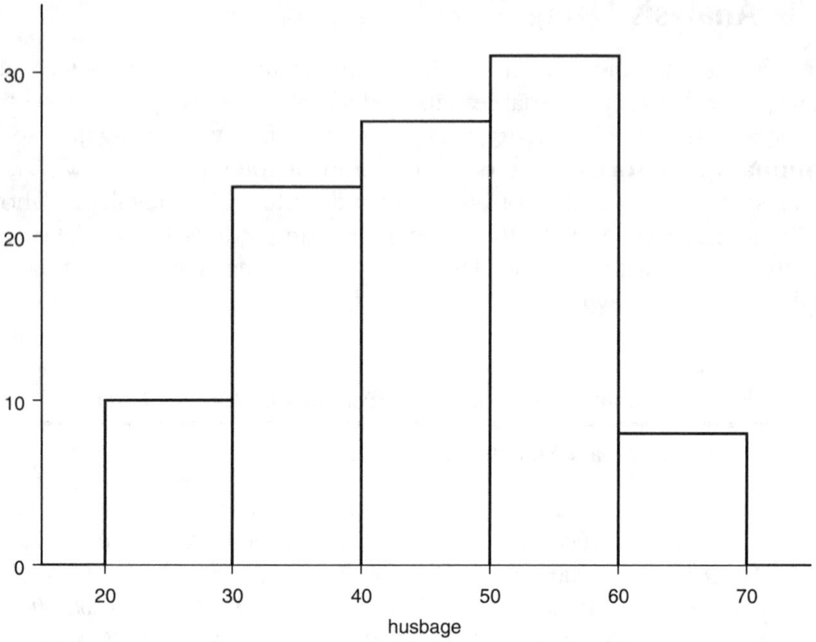

Figure 2.1   Histogram of husband's age.

In addition to the basic numerical summaries given in Table 2.2, graphics are an essential part of the initial examination of data. Here we shall begin by constructing histograms for each of the five variables first by using the **Graphics menu**.

■ Click on **Graph**.
■ Select **2D** and then highlight **Histogram (*x*)**.
■ Click **OK** and the **Histogram/Density** dialog appears.
■ Select **huswif** as the data set and say **husbage** as the *x* column.
■ Click **OK** to give Figure 2.1, a histogram of husband's age.

A similar sequence of instructions can be used to produce histograms of each of the other variables, and by repeated use of **Insert, Graph**, they can all be arranged on the same sheet as shown in Figure 2.2. (On the screen the diagram will be colour; to print on a black and white printer use **Format, Apply Style, Black and White** before **File, Print Graph**.)

More useful than Figure 2.2 would be diagrams that contain a number of alternative graphical displays of the same variable, for example, a histogram, a box plot, and a normal probability plot. We now examine how this can be constructed for the **husbage** variable using the command language.

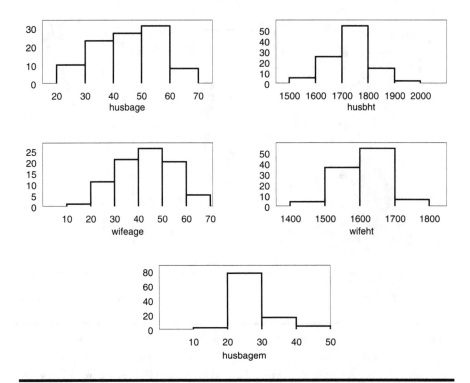

Figure 2.2 **Histograms of all five variables in the** huswif **data frame.**

```
attach(huswif)
#set up plotting area to take three graphs
par(mfrow=c(1,3))
#use hist function to plot histogram and label x and y
#axis appropriately
hist(husbage,xlab="Age of husband",ylab="Frequency")
#
#use boxplot function and label with variable information
boxplot(husbage,ylab="Age of husband")
#use qqnorm function to construct a normal probability
#plot
qqnorm(husbage)
qqline(husbage)
```

The resulting diagram is shown in Figure 2.3. The box plot shows that there are no obvious outliers on this variable, and the histogram suggests some degree of non-normality, also indicated in the probability plot. Similar diagrams could be drawn for the other four variables. We leave

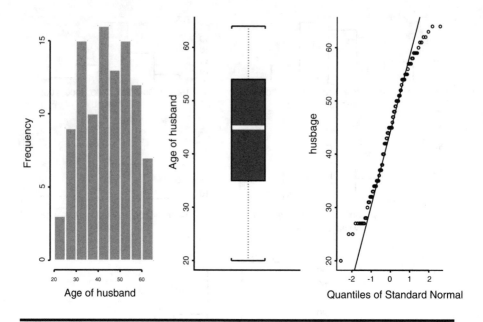

Figure 2.3   Histogram, box plot, and probability plot of husband's age.

this and the construction of Figure 2.3 using the GUI as exercises for the reader.

Having obtained both numerical and graphical information about individual variables in the data, we might now want to look at the relationships between variables. The most generally used numerical summary of these relationships is the *correlation matrix* of the variables. This can be obtained via the S-PLUS GUI as follows:

■ Click on **Statistics**.
■ Select **Data Summaries**.
■ Select **Correlations**.

In the **Correlations and Covariance** dialog first select the huswif data set; **<ALL>** is highlighted by default in the variables section of the dialog, meaning that all variables in the chosen data frame will be included in the calculation of the correlation matrix. We now need to consider what to select in the **Method to Handle Missing Values** box. The default of **Fail** needs to be changed for the huswif data set. So select **Available** and then click **OK** to give the correlation matrix shown in Table 2.3. The correlations in this matrix are calculated on the basis of all observations available for each pair of variables.

Table 2.3 Correlation Matrix for Husbands and Wives Data

*** Correlations for data in: huswif ***

	husbage	husbht	wifeage	wifeht	husbagem
husbage	1.0000000	–0.20018660	0.96749749	–0.2358639	0.3002488
husbht	–0.2001866	1.00000000	–0.01332892	0.3594675	–0.2575508
wifeage	0.9674975	–0.01332892	1.00000000	–0.2259641	0.1118882
wifeht	–0.2358639	0.35946747	–0.22596407	1.0000000	–0.1402032
husbagem	0.3002488	–0.25755083	0.11188819	–0.1402032	1.0000000

The values in the correlation matrix show that the ages of husbands and their wives are highly correlated (0.95); other pairs of variables, however, have only moderate correlations.

With the command language, the same correlation matrix can be found using the **cor** function:

```
cor(huswif,na.method="available")
#the na.method argument is used to select a procedure for
#dealing with missing values
```

Assessing the relationships between variables simply on the basis of the numerical values of their correlations is not, in general, to be recommended. Correlations can be badly distorted, for example, by outliers in the data, and can give misleading values if the relationships between the variables are anything but linear. Consequently, it is important to look at the numerical correlations alongside scatterplots of the variables. Scatterplots of each pair of variables are obtained either via the **Graphics menu** or the **plot** command, both of which were illustrated in Chapter 1. Here there are ten possible scatterplots for the five variables, and it is convenient to view them displayed in the form of what is usually called a *scatterplot matrix*. Formally, a scatterplot matrix is defined as a square symmetric grid of bivariate scatterplots (Cleveland, 1993). The grid has p rows and p columns, each one corresponding to a different one of the p observed variables. Each of the cells of the grid shows a scatterplot of two variables. Because the scatterplot is symmetric about its diagonal, variable i is plotted against variable j in the ijth cell, and the same variables also appear in cell ji with the x and y axes of the scatterplot interchanged. The reason for including both the upper and lower triangles in the matrix, despite the seeming redundancy, is that it enables a row or column to be visually scanned to see one variable against all others, with the scale for one variable lined up along the horizontal or the vertical.

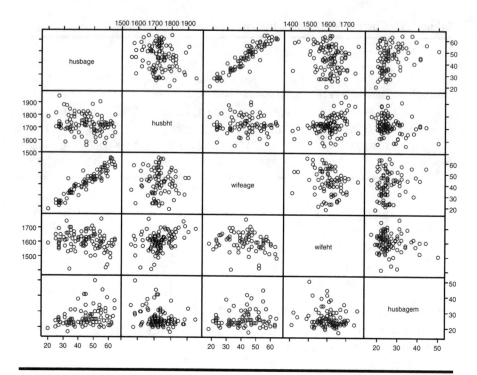

Figure 2.4   Scatterplot matrix for all five variables in the huswif **data frame.**

To obtain the scatterplot matrix of the five variables in the huswif data frame we can use the **Graphics menu** as follows:

- ■ Click **Graph**.
- ■ Choose **2D** and under **Axes Type** highlight **Matrix**.
- ■ Click on **OK** and the **Scatter Plot Matrix** dialog appears.
- ■ Select the **huswif** data frame and highlight all variables in the *x* columns box.
- ■ Click **OK**.

This leads to the scatterplot matrix seen in Figure 2.4

The pattern of the relationships between the pairs of variables is made very clear in this diagram and it is also reassuring to see that there are no very obvious outliers or nonlinear relations with which to be concerned. Consequently, the correlation matrix in Table 2.3 does represent a reasonable numerical summary of the relationships between the five variables in the data.

With the command language, the scatterplot matrix can be obtained with the **pairs** function as follows:

```
pairs(huswif)
```

In some cases it may be informative to label the points in a scatterplot (and in a scatterplot matrix) in some way. Suppose for the **huswif** data we would like to label points according to whether the husband or wife is tallest, say "hw" if the husband is taller and "WH" if the wife is taller. The following commands label the points in this way in a scatterplot matrix of the data using the **panel=function** argument to do something other than simply the default, which is to plot the points in each panel of the grid of scatterplots.

```
difht<-husbht-wifeht
#find difference in heights of the couples
labs<-rep("hw",100)
#use the rep function to create a vector
#of length 100, each element of which is the label hw
labs[difht<=0]<-"WH"
#change elements of labs to WH for those
#couples where the wife is taller or equal
#in height to her husband
pairs(huswif,panel=function(x,y)text(x,y,labels=labs,cex=0.5))
#use the text function to plot the contents
#of labs in the appropriate positions in each
#scatterplot; cex is a parameter controlling
#the size of the plotted character.
```

The resulting plot is shown in Figure 2.5. As might be expected, most points are labelled 'hw', since husbands are, in general, taller than their wives. (Many other examples of using the **panel=function** argument of the **pairs** function will be given in later chapters.)

Earlier in this chapter probability plots were used to assess individual variables in the **huswif** data frame for normality. But it may be as important to assess the five variables jointly for multivariate normality. One possible method that can be used is a *chi-squared probability plot* of the *generalized distances* (*Mahalanobis distance*) of each observation from the mean vector of the data, i.e., for observation i with vector of values $\mathbf{x}_i$ the distance d_i given by

$$d_i = \left(\mathbf{x}_i - \bar{\mathbf{x}}\right)' \mathbf{S}^{-1}\left(\mathbf{x}_i - \bar{\mathbf{x}}\right) \tag{2.1}$$

where $\bar{\mathbf{x}}$ is the sample mean vector and $\mathbf{S}$ the sample covariance matrix. If the observations do arise from a multivariate normal distribution, these distances have, approximately, a chi-squared distribution with degrees of

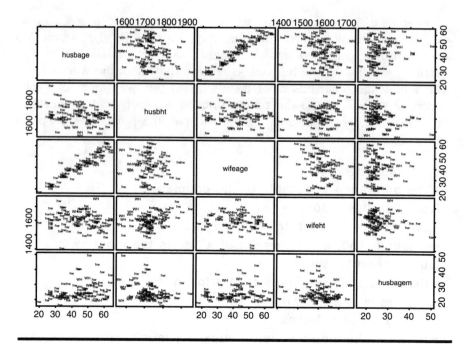

Figure 2.5   Scatterplot matrix for all five variables in the huswif **data frame with points labelled according to whether husband (hw) or wife (WH) is taller.**

freedom equal to the number of variables. So plotting the ordered distances against the corresponding quantiles of the appropriate chi-squared distribution should lead to a straight line through the origin. Such a plot is easily constructed using the S-PLUS command language and in particular the **Mahalanobis** function; suitable code is as follows;

```
#first use the na.exclude function to remove observations
#with missing values on any variable
huswif1<-na.exclude(huswif)
#use the apply function to find the mean vector of the
#data
meanv<-apply(huswif1,2,mean)
#use var function to find covariance matrix
S<-var(huswif1)
#use length function to find number of
#observations in huswif1
n<-length(huswif1[,1])
#Get sample quantiles
index <-(1:n)/(n+1)
```

```
#1:n produces the vector of values 1,2,...,n
#get corresponding chi-squared quantiles
quant <-qchisq(index,5)
#get generalized distances using Mahalanobis
#function
dist<-mahal(huswif1,mean,S)
#dist will contain the Mahalanobis distances of each
#observation from the mean vector of the data.
#plot ordered distances obtained using the
#sort function
plot(quant,sort(dist),ylab="Ordered distances",
xlab="Chi-square quantiles",pch= 1)
abline(0,1)
#use the abline function to add a line with intercept 0
#and slope 1, i.e., the line y=x
```

The resulting plot is shown in Figure 2.6. There is considerable deviation from a straight line in the plotted points. This gives relatively clear evidence that the data do not have a multivariate normal distribution, which may have implications for particular types of analysis that may be considered for the data.

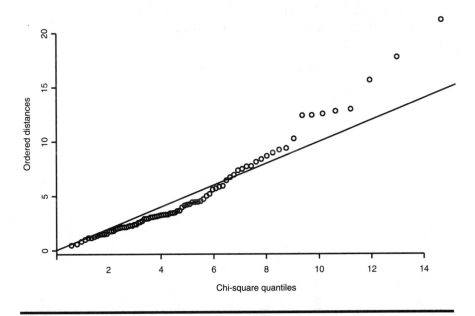

Figure 2.6   Chi-squared probability plot of all five variables in the huswif data frame.

Exercises

2.1. It is often important to include information about the marginal distributions of each variable on a scatterplot of a pair of variables. A simple procedure for this is the use of the **rug** function in S-PLUS (with perhaps the **jitter** function as well). Investigate by producing a scatterplot of husband's age against wife's age showing the marginal distributions of each variable.

2.2. Investigate the use of the **density** function in S-PLUS for enhancing histograms with a nonparametric estimate of the distribution of the variables. Additionally, examine the use of both the **Graph menu** and the **2D graphical palette** for the same thing.

2.3. Produce a scatterplot matrix of the variables in the husbands/wives data, with each panel the value of the correlation coefficient of the two variables shown.

2.4. Use the chi-squared plotting technique for assessing multivariate normality described in the text to assess only the first four variables in the **huswif** data frame. Do you think these four variables do have a multivariate normal distribution?

2.5. Explore the use of the **hist2d** function to get two-dimensional histograms for husband's age and wife's age. Use the **persp** function to view the histogram.

2.6. Use the **Scatter Plot Matrix** dialog to get the scatterplot matrix of the five variables in the **huswif** data frame, enhanced with a histogram of each variable on the main diagonal.

Chapter 3

Analysis of Variance: Poisoning Rats and Losing Weight

3.1 Description of Data

In this chapter we shall consider the analysis of two sets of data. The first, shown in Table 3.1, involves the survival times (in 10-hour units) of rats subjected to three different poisons and four different treatments. This is an example of a 3×4 factorial design with four replications in each cell.

The second data set we shall be concerned with here is shown in Table 3.2. These data arise from a study into the effectiveness of slimming clinics. Of particular interest is the question of whether adding a technical manual giving advice based on psychological behaviourist theory to the support offered would help clients to control their diet. So the first factor is Condition with two levels: 'Experimental' (those given the manual) and 'Control' (those without the manual). It was also thought important to distinguish between clients who had already been trying to slim and those who had not, giving rise to the second factor, Status, again with two levels: 'Experienced' (those who had been trying to slim for more than 1 year) and 'Inexperienced' (those who had been trying to slim for not more than 3 weeks). The response variable recorded for each participant in the study was

$$\frac{\text{weight at 3 months} - \text{ideal weight}}{\text{Initial weight} - \text{ideal weight}} \tag{3.1}$$

Table 3.1   Survival Times of Rats

Poison	Treatment	Time
1	1	0.31
1	1	0.45
1	1	0.46
1	1	0.43
1	2	0.82
1	2	1.10
1	2	0.88
1	2	0.72
1	3	0.43
1	3	0.45
1	3	0.63
1	3	0.76
1	4	0.45
1	4	0.71
1	4	0.66
1	4	0.62
2	1	0.36
2	1	0.29
2	1	0.40
2	1	0.23
2	2	0.92
2	2	0.61
2	2	0.49
2	2	1.24
2	3	0.44
2	3	0.35
2	3	0.31
2	3	0.40
2	4	0.56
2	4	1.02
2	4	0.71
2	4	0.38
3	1	0.22
3	1	0.21
3	1	0.18
3	1	0.23
3	2	0.30
3	2	0.37
3	2	0.38
3	2	0.29

Table 3.1 (Continued) Survival Times of Rats

Poison	Treatment	Time
3	3	0.23
3	3	0.25
3	3	0.24
3	3	0.22
3	4	0.30
3	4	0.36
3	4	0.31
3	4	0.33

In this case the design is a 2 × 2 factorial, but the number of observations in each of the four cells is not the same; the data are *unbalanced*, rather than *balanced* as in the first example.

3.2 Analysis of Variance

Analysis of variance (ANOVA) encompasses a set of methods for testing hypotheses about differences between means. Underlying ANOVA is a linear model, which for both data sets introduced in the previous section would have the following form:

$$y_{ijk} = \mu + \alpha_i + \beta_j + \gamma_{ij} + \epsilon_{ijk} \qquad (3.2)$$

where y_{ijk} represents the k observation in the jth level of one factor and the ith level of the other. The parameters μ, α_i, β_j, and γ_{ij} represent, respectively, overall mean, main effects of each factor, and the interaction between the two factors. The ϵ_{ijk} are error terms assumed to have a normal distribution with zero mean and variance σ^2. (To make the model in Equation 3.2 identifiable, the parameters need to be constrained in some way, for example, by requiring that their sum over any subscript is zero.)

For the balanced survival time data, the variability in the observations can be portioned into sums of squares representing poisons, treatments, and the poisons × treatments interaction, *orthogonally*; i.e., they would be nonoverlapping. But for the unbalanced slimming data, this is not the case, because, for example, there is a proportion of the variability in the response variable that can be attributed to (explained by) either of the two factors in the study. A consequence is that Condition and Status

Table 3.2 Slimming Data

Condition	Status	Response
1	1	−14.67
1	1	−1.85
1	1	−8.55
1	1	−23.03
1	1	11.61
1	2	0.81
1	2	2.38
1	2	2.74
1	2	3.36
1	2	2.10
1	2	−0.83
1	2	−3.05
1	2	−5.98
1	2	−3.64
1	2	−7.38
1	2	−3.60
1	2	−0.94
2	1	−3.39
2	1	−4.00
2	1	−2.31
2	1	−3.60
2	1	−7.69
2	1	−13.92
2	1	−7.64
2	1	−7.59
2	1	−1.62
2	1	−12.21
2	1	−8.85
2	2	5.84
2	2	1.71
2	2	−4.10
2	2	−5.19
2	2	0.00
2	2	−2.80

together explain less of the variation in the response variable than the sum of which each explains alone. The result is that the sum of squares that can be attributed to a factor depends on which other factors have already been allocated a sum of squares; in other words, the sums of squares of the factors now depend on the order in which they are considered. There is a lack of uniqueness in partitioning the variation in the response variable. This is not so for balanced data. (For a more detailed discussion of the problems with unbalanced designs, see Nelder, 1977, and Aitkin, 1978).

3.3 Analysis Using S-PLUS

The data sets in Tables 3.1 and 3.2 are available as S-PLUS data frames called **rats** and **slim**, respectively, with the factors appropriately defined as S-PLUS factor variables and their levels suitably labelled.

In this chapter we shall be applying the analysis of variance model specified in Equation 3.2 and so this is a convenient point to say a little about the model formulae used in S-PLUS which are common to all its modelling procedures. A formula in S-PLUS is a symbolic expression that defines the structural form of the model and is interpreted by modelling functions. A simple example of an S-PLUS model formula is

```
y~x1+x2+x3
```

y is the response variable and *x*1 to *x*3 explanatory variables. The corresponding regression coefficients and the constant are implied in the formula. The term on the left-hand side can be an expression representing, say, a transformation of the response variable, for example,

```
log(y)~x1+x2+x3
```

A model formula such as

```
y~x1*x2
```

means that *y* is modelled using the main effects *and* the interaction of *x*1 and *x*3. Written out in full the model is

```
y~x1+x2+x1:x2
```

with the colon used to indicate the interaction term.

Variables on the right-hand side of a model formula can also be transformed, but since arithmetic operators have a different meaning in formulae (as we have seen above), the expressions must be enclosed in the **I** function; for example,

```
y~I(x1*x2)
```

means that *y* is modelled in terms of a single explanatory variable given by the product of *x1* and *x2*, in addition to the constant.

More aspects of model formulae will be introduced in later chapters.

3.3.1 *Analysis of Variance of Survival Times of Rats*

The contents of the **rats** data frame are shown in Table 3.3. To begin, let us simply obtain the analysis of variance table for these data using the appropriate dialog box:

- Click on **Statistics**.
- Select **ANOVA**.
- Select **Fixed Effects**.

The **ANOVA** dialog now becomes visible, and we proceed as follows:

- Select the **rats** data frame.
- Click on the **Create Formula** tab to access the **Formula** dialog.
- Highlight **Time** in the **Choose Variables Section**.
- Click on the **Response** tab.

The following now appears in the **Formula** section.

    Time~1

- Highlight both **Poison** and **Treatment** and check the **Main + Interact** tab to give the following formula:

    Time~Poison*Treatment

This corresponds to the required main effects plus interaction model.

- Click on **OK** to return to the **ANOVA** dialog.
- Click **OK**.

This leads to the results shown in Table 3.4. The analysis of variance table indicates that the Poison × Treatment interaction is nonsignificant, but that both Poison and Treatment main effects are significant. (Note that the S-PLUS commands equivalent to the analysis carried out using the dialog approach appear in the Report file.)

So it appears that a simple main effects model is suitable for these data. We might now move on to consider the following:

Table 3.3   Contents of the rats Data Frame

> rats

	Poison	Treatment	Time
1	P1	A	0.31
2	P1	A	0.45
3	P1	A	0.46
4	P1	A	0.43
5	P1	B	0.82
6	P1	B	1.10
7	P1	B	0.88
8	P1	B	0.72
9	P1	C	0.43
10	P1	C	0.45
11	P1	C	0.63
12	P1	C	0.76
13	P1	D	0.45
14	P1	D	0.71
15	P1	D	0.66
16	P1	D	0.62
17	P2	A	0.36
18	P2	A	0.29
19	P2	A	0.40
20	P2	A	0.23
21	P2	B	0.92
22	P2	B	0.61
23	P2	B	0.49
24	P2	B	1.24
25	P2	C	0.44
26	P2	C	0.35
27	P2	C	0.31
28	P2	C	0.40
29	P2	D	0.56
30	P2	D	1.02
31	P2	D	0.71
32	P2	D	0.38
33	P3	A	0.22
34	P3	A	0.21
35	P3	A	0.18
36	P3	A	0.23
37	P3	B	0.30
38	P3	B	0.37

Table 3.3 (Continued) Contents of the rats Data Frame

> rats

	Poison	Treatment	Time
39	P3	B	0.38
40	P3	B	0.29
41	P3	C	0.23
42	P3	C	0.25
43	P3	C	0.24
44	P3	C	0.22
45	P3	D	0.30
46	P3	D	0.36
47	P3	D	0.31
48	P3	D	0.33

Table 3.4 Analysis of Variance of Survival Times of Rats

*** Analysis of Variance Model ***

Short Output:
Call:
  aov(formula = Time ~ Poison * Treatment, data = rats,
  na.action = na.exclude)

Terms:

	Poison	Treatment	Poison:Treatment	Residuals
Sum of Squares	1.033	0.921	0.250	0.801
Deg. of Freedom	2	3	6	36

Residual standard error: 0.1491
Estimated effects are balanced

	Df	Sum of Sq	Mean Sq	F Value	Pr(F)
Poison	2	1.033	0.5165	23.22	0.0000
Treatment	3	0.921	0.3071	13.81	0.0000
Poison:Treatment	6	0.250	0.0417	1.87	0.1123
Residuals	36	0.801	0.0222		

1. Whether the normality assumption made by the analysis of variance model is justified for these data;
2. Whether the constant variance assumption is justified;
3. Use of *multiple comparison tests* to examine in more detail which Poison means and which Treatment means differ.

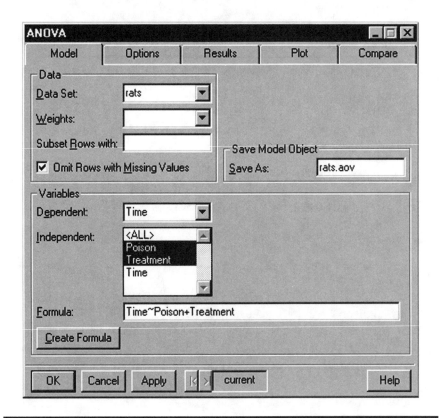

Figure 3.1   ANOVA dialog showing main effects model for the data on survival times of rats.

To check both the normality and constant variance assumptions, we shall examine the *residuals* from the fitted model, i.e., the differences between the observed values and those predicted by the model. A normal probability plot of the residuals will be used to assess assumption (1), and a plot of residuals against fitted values can be used to evaluate assumption (2). Departures from a 'horizontal band' shape give cause for concern. (More will be said about the use of residuals for diagnosing models in later chapters.)

So we return to the **ANOVA** dialog, set up the main effects model for the rats data frame, and request that the resulting model object be saved as, say, rats.aov (this is needed for the multiple comparison tests to be described later). The **ANOVA** dialog now appears as shown in Figure 3.1.

Now click on the **Plot** tab and tick the **Residuals v Fit** and **Residuals Normal QQ** options. Finally click on **OK** to give the results shown in Table 3.5 and the plots shown in Figures 3.2 and 3.3.

Table 3.5 Analysis of Variance of Survival Times of Rats: Main Effects Model

*** Analysis of Variance Model ***

Short Output:
Call:
    aov(formula = Time ~ Poison + Treatment, data = rats,
        na.action = na.exclude)

	Terms		
	Poison	Treatment	Residuals
Sum of Squares	1.033	0.921	1.051
Deg. of Freedom	2	3	42

Residual standard error: 0.1582
Estimated effects are balanced

	Df	Sum of Sq	Mean Sq	F Value	Pr(F)
Poison	2	1.033	0.5165	20.64	5.700e-007
Treatment	3	0.921	0.3071	12.27	6.697e-006
Residuals	42	1.051	0.0250		

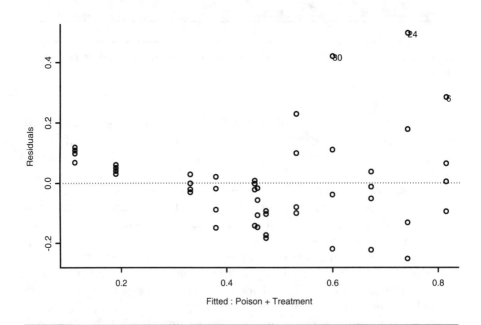

Figure 3.2   Residuals from main effects model for survival times of rats plotted against fitted values.

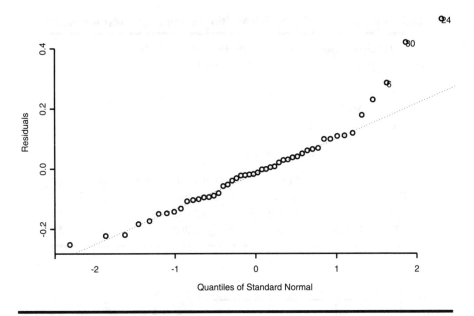

Figure 3.3 Normal probability plot of residuals from main effects model fitted to rat survival data.

The analysis of variance table in Table 3.5 confirms the highly significant main effects of both Poison and Treatment. The normal plot of the residuals suggests some departure from normality and the residuals vs. fitted values plot has a 'wedge shape' consistent with a departure from the constant variance assumption. Both findings suggest that analysing the survival times after a suitable transformation might be more appropriate than analysing the raw data. We leave this possibility to be investigated by the reader (see Exercise 3.1), and move on here to consider the use of multiple comparison tests to assess just which poison means and which treatment means differ. We shall use a test due to Scheffé. (This test in particular and multiple comparison tests in general are described in Everitt, 2001.)

- Click on **Statistics**.
- Select **ANOVA**.
- Select **Multiple Comparisons**.

In the **Multiple Comparisons** dialog, select the rats.aov created by previous use of the **ANOVA** dialog, select **Poison** in the **Levels of** section, and in the Method section select **Scheffé**. Click **OK** to see the numerical results in Table 3.6 and the graphical display of these results as seen in Figure 3.4. The results indicate that the mean survival time for P3 is

Table 3.6   Results of Scheffé's Multiple Comparison Test for Poisons

95% simultaneous confidence intervals for specified
linear combinations, by the Scheffé method

critical point: 2.5377
response variable: Time
rank used for Scheffé method: 2

intervals excluding 0 are flagged by '****'

	Estimate	Std.Error	Lower Bound	Upper Bound	
P1-P2	0.0731	0.0559	–0.0688	0.215	
P1-P3	0.3410	0.0559	0.1990	0.483	****
P2-P3	0.2680	0.0559	0.1260	0.410	****

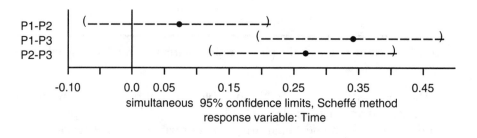

Figure 3.4   Scheffé multiple comparison results for levels of Poison in rat survival data.

significantly less than for P1 and P2, which themselves do not differ in mean survival time.

Repeating the process but now selecting **Treatment** leads to the results in Table 3.7 and Figure 3.5. It appears that the mean survival times for treatments A and B, A and D, and B and C differ.

3.3.2   Analysis of Variance of Slimming Data

The contents of the **slim** data frame are shown in Table 3.8. To analyse this data set we shall use the command language approach. The unbalanced nature of the data can be seen by examining the cell counts in the design. These are obtained using the **tapply** function and the **length** function

```
tapply(Response,list(Status,Condition),length)
```

Table 3.7 Results of Scheffé's Test for Treatments

95% simultaneous confidence intervals for specified linear combinations, by the Scheffé method

critical point: 2.9122
response variable: Time
rank used for Scheffé method: 3

intervals excluding 0 are flagged by '****'

	Estimate	Std.Error	Lower Bound	Upper Bound	
A-B	–0.3620	0.0646	–0.5510	–0.1740	****
A-C	–0.0783	0.0646	–0.2660	0.1100	
A-D	–0.2200	0.0646	–0.4080	–0.0319	****
B-C	0.2840	0.0646	0.0961	0.4720	****
B-D	0.1420	0.0646	–0.0456	0.3310	
C-D	–0.1420	0.0646	–0.3300	0.0464	

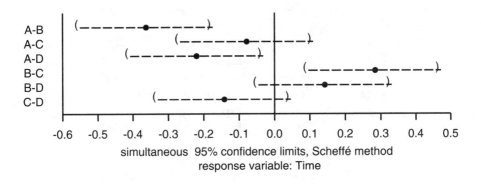

Figure 3.5 Scheffé multiple comparison results for Treatment levels in rat survival data.

to give the following:

	Experimental	Control
Experienced	5	11
Inexperienced	12	6

To begin, it may be helpful to examine some simple plots of the data. A plot of mean values for each level of each factor compared to a corresponding

Table 3.8 Contents of the slim Data Frame

	Condition	Status	Response
1	Experimental	Experienced	−14.67
2	Experimental	Experienced	−1.85
3	Experimental	Experienced	−8.55
4	Experimental	Experienced	−23.03
5	Experimental	Experienced	11.61
6	Experimental	Inexperienced	0.81
7	Experimental	Inexperienced	2.38
8	Experimental	Inexperienced	2.74
9	Experimental	Inexperienced	3.36
10	Experimental	Inexperienced	2.10
11	Experimental	Inexperienced	−0.83
12	Experimental	Inexperienced	−3.05
13	Experimental	Inexperienced	−5.98
14	Experimental	Inexperienced	−3.64
15	Experimental	Inexperienced	−7.38
16	Experimental	Inexperienced	−3.60
17	Experimental	Inexperienced	−0.94
18	Control	Experienced	−3.39
19	Control	Experienced	−4.00
20	Control	Experienced	−2.31
21	Control	Experienced	−3.60
22	Control	Experienced	−7.69
23	Control	Experienced	−13.92
24	Control	Experienced	−7.64
25	Control	Experienced	−7.59
26	Control	Experienced	−1.62
27	Control	Experienced	−12.21
28	Control	Experienced	−8.85
29	Control	Inexperienced	5.84
30	Control	Inexperienced	1.71
31	Control	Inexperienced	−4.10
32	Control	Inexperienced	−5.19
33	Control	Inexperienced	0.00
34	Control	Inexperienced	−2.80

plot of medians is sometimes a good place to start because it may highlight possible distribution problems with the data. The required plots may be constructed using the **plot.design** function.

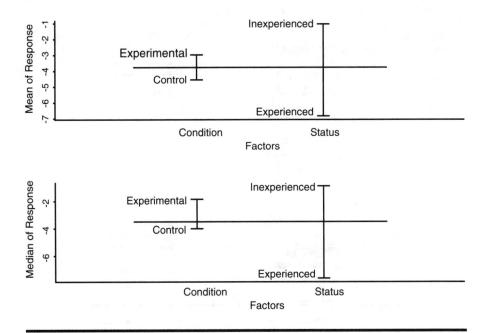

Figure 3.6   Plots of means and medians for slimming data.

```
par(mfrow=c(2,1))
#sets up plotting area to take both mean and
#median plots
plot.design(slim)
plot.design(slim,fun=median)
#fun=median specifies that the medians are to
#be plotted; the default is the mean
```

The resulting diagram is shown in Figure 3.6. There is some difference between treatment means and medians, particularly for the Conditions factor. This may result from one or two outliers and we can examine this possibility with a further simple diagram, namely, a box plot of the observations at each level of each factor; this is constructed using the **plot.factor** function.

```
par(mfrow=c(2,1))
plot.factor(slim)
```

The resulting diagram appears in Figure 3.7. A number of outliers are revealed in this diagram. There is, for example, one participant in the experimental group with a very large decrease in weight (–22.03) and in

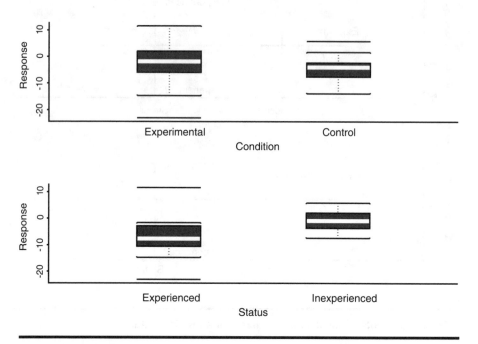

Figure 3.7   Box plots for slimming data.

the 'experienced' slimmers group there is one person with a large increase in weight (11.61). Consideration might be given to removing these observations but here we shall continue to include them in our analyses, and leave the reader to investigate the former possibility (see Exercise 3.2).

Now we will use the **aov** function to fit various analysis of variance models to the data, beginning with models including only the main effect of Condition and then the main effects of both Condition and Status.

```
summary(aov(Response~Condition,data=slim))
# the aov function requires a valid S-PLUS
# formula and, if not already attached,
# the name of a data frame.
# aov returns an analysis of variance object
# which when given to the summary function
# produces an analysis of variance table etc
summary(aov(Response~Condition+Status,data=slim))
```

The results are shown in Table 3.9. Now rerun, taking Status first followed by Condition (we now assume that the **slim** data frame is attached).

Table 3.9   Analyses of Variance of the Slimming Data with Status Added after Condition

```
> summary(aov(Response ~ Condition))
```

	Df	Sum of Sq	Mean Sq	F Value	Pr(F)
Condition	1	21	21.19	0.5041	0.4828
Residuals	32	1345	42.03		

```
> summary(aov(Response ~ Condition + Status))
```

	Df	Sum of Sq	Mean Sq	F Value	Pr(F)
Condition	1	21	21.2	0.609	0.4412
Status	1	266	265.9	7.640	0.0095
Residuals	31	1079	34.8		

Table 3.10   Analyses of Variance of the Slimming Data with Condition Added after Status

```
> summary(aov(Response ~ Status))
```

	Df	Sum of Sq	Mean Sq	F Value	Pr(F)
Status	1	285	285.0	8.435	0.006621
Residuals	32	1081	33.8		

```
> summary(aov(Response ~ Status + Condition))
```

	Df	Sum of Sq	Mean Sq	F Value	Pr(F)
Status	1	285	285.0	8.187	0.0075
Condition	1	2	2.1	0.061	0.8062
Residuals	31	1079	34.8		

```
summary(aov(Response~Status))
summary(aov(Response~Status+Condition))
```

The results of these two commands are shown in Table 3.10. Notice how the sums of squares for each factor differ depending on whether they appear first or second in the model. This would not be the case if the design were balanced (readers can confirm this by using the series of commands above, suitably amended, to analyse the data in the **rats** data frame).

Finally, we can fit a model containing both main effects and the interaction,

```
summary(aov(Response~Status*Condition))
```

to give the results shown in Table 3.11. Clearly, the interaction is not significant, and the earlier results show that the levels of the Condition

Table 3.11   Analysis of Variance of Slimming Data

```
> summary(aov(Response ~ Status + Condition + Status:
        Condition))
```

	Df	Sum of Sq	Mean Sq	F Value	Pr(F)
Status	1	285	285.0	7.924	0.0085
Condition	1	2	2.1	0.059	0.8094
Status:Condition	1	0	0.1	0.004	0.9524
Residuals	30	1079	36.0		

factor do not have different means, but that there is a significant difference in the means of Experienced and Inexperienced slimmers. The two means can be found by again using the **tapply** function.

```
tapply(Response,list(Status),mean)
```

giving

```
Experienced    Inexperienced

   −6.832           −1.032
```

Experienced slimmers show, on average, a greater weight decrease than inexperienced slimmers.

Exercises

3.1.   Reanalyse the survival times of rats data after taking a log transformation. Examine the residuals from whatever model you now find
is appropriate to assess whether the normality and constant variance
assumptions are met more satisfactorily than when modelling the
raw data.

3.2.   Reanalyse the slimming data after removing the two possible outliers
identified in the text. Are the conclusions from the analysis the same
as those discussed in the text?

3.3.   Use the **predict** function on the results of a main effects analysis of
variance model for the survival time data to find the fitted values.
What do these fitted values correspond to in terms of Poison and
Treatment means?

3.4.   Investigate what happens if the **aov** function is applied to the
slimming data with the Status × Condition interaction listed before
the main effects of the two factors.

Chapter 4

Multiple Regression: Technological Change in Jet Fighters

4.1 Description of Data

The data to be used in this chapter were introduced previously in Chapter 1; see Table 1.1. The data give the values of six variables for 22 U.S. fighter aircraft. The variables are as follows:

FFD: First flight date, in months after January 1940
SPR: Specific power, proportional to power per unit weight
RGF: Flight range factor
PLF: Payload as a fraction of gross weight of aircraft
SLF: Sustained load factor
CAR: A binary variable specifying whether the aircraft can or cannot land on a carrier

Interest lies in modelling FFD as a function of the other variables. Here we shall concentrate on using *multiple regression*.

4.2 Multiple Regression Model

The multiple regression model has the general form

$$y_i = \beta_0 + \beta_1 x_{i1} + \beta_2 x_{i2} + \beta_p x_{ip} + \epsilon_i \qquad (4.1)$$

where y_i and x_{i1}, x_{i2} ... x_{ip} are, respectively, the values of a response variable and p explanatory variables for the ith observation in a sample of size n. The *regression coefficients,* β_0, β_1 ... β_p are generally estimated by least squares — they represent the expected change in the response variable predicted by a unit change in the corresponding explanatory variables conditional on the values of the remaining explanatory variables. Significance tests for the regression coefficient can be derived by assuming that the residual terms ϵ_i, $i = 1$ ... n, are from a normal distribution with zero mean and constant variance σ^2.

For n observations of the response and explanatory variables, the regression model may be written concisely as

$$y = X\beta + \epsilon \qquad (4.2)$$

where y is the $n \times 1$ vector of responses, X is an $n \times (p + 1)$ matrix of known constants, the first column containing a series of ones corresponding to the term β_0 in Equation 4.1 and the remaining columns values of the explanatory variables. The elements of the vector β are the regression coefficients β_0 ... β_p and those of the vector ϵ, the residual terms ϵ_1 ... ϵ_n.

Full details of multiple regression are given in Rawlings (1988).

4.3 Analysis Using S-PLUS

The data are stored in the data frame **jets**, the contents of which are shown in Table 4.1.

Before proceeding with the regression modelling of the data it may be helpful to examine a scatterplot matrix of the variables, particularly if we label the points by aircraft type. By using the GUI, this involves the following:

- Click **Graph**.
- Select **2D**.
- Highlight **Matrix** under **Axes Type**, and click **OK**.
- Select the **jets** data set.
- In the *x* columns box highlight all variables but **Type** and **CAR**.
- In the *y* columns box highlight **Type**.

Table 4.1   The jets Data Frame

	Type	FFD	SPR	RGF	PLF	SLF	CAR
1	FH-1	82	1.468	3.30	0.166	0.10	Cannot land
2	FJ-1	89	1.605	3.64	0.154	0.10	Cannot land
3	F-86A	101	2.168	4.87	0.177	2.90	Can land
4	F9F-2	107	2.054	4.72	0.275	1.10	Cannot land
5	F-94A	115	2.467	4.11	0.298	1.00	Can land
6	F3D-1	122	1.294	3.75	0.150	0.90	Cannot land
7	F-89A	127	2.183	3.97	0.000	2.40	Can land
8	XF10F-1	137	2.426	4.65	0.117	1.80	Cannot land
9	F9F-6	147	2.607	3.84	0.155	2.30	Cannot land
10	F-100A	166	4.567	4.92	0.138	3.20	Can land
11	F4D-1	174	4.588	3.82	0.249	3.50	Cannot land
12	F1F-1	175	3.618	4.32	0.143	2.80	Cannot land
13	F-101A	177	5.855	4.53	0.172	2.50	Can land
14	F3H-2	184	2.898	4.48	0.178	3.00	Cannot land
15	F-102A	187	3.880	5.39	0.101	3.00	Can land
16	F-8A	189	0.455	4.99	0.008	2.64	Cannot land
17	F-104B	194	8.088	4.50	0.251	2.70	Can land
18	F-105B	197	6.502	5.20	0.366	2.90	Can land
19	YF-107A	201	6.081	5.65	0.106	2.90	Can land
20	F-106A	204	7.105	5.40	0.089	3.20	Can land
21	F-4B	255	8.548	4.20	0.222	2.90	Cannot land
22	F-111A	328	6.321	6.45	0.187	2.00	Can land

■ Click on the **Symbol** tab and tick the **Use Text As Symbol** box.
■ Specify *y* column as **Text to Use**.
■ Click **OK**.

This leads to the labelled scatterplot matrix seen in Figure 4.1. In some panels fighter F-111A seems to be a possible outlier, as, to a lesser extent, does the F-105B. But at this stage it seems unwise to remove them, particularly given the small sample size.

Some of the panels in Figure 4.1 suggest perhaps that the relationship between some pairs of variables, at least, are nonlinear, for example, SPR and SLF. To investigate this possibility further before undertaking formal modelling of the data, we can plot the simple regression line linking each pair of variables onto the appropriate panel and a *locally weighted regression fit* (see Chambers and Hastie, 1992, for comparison). To avoid cluttering each scatterplot in the diagram we shall in this case not label

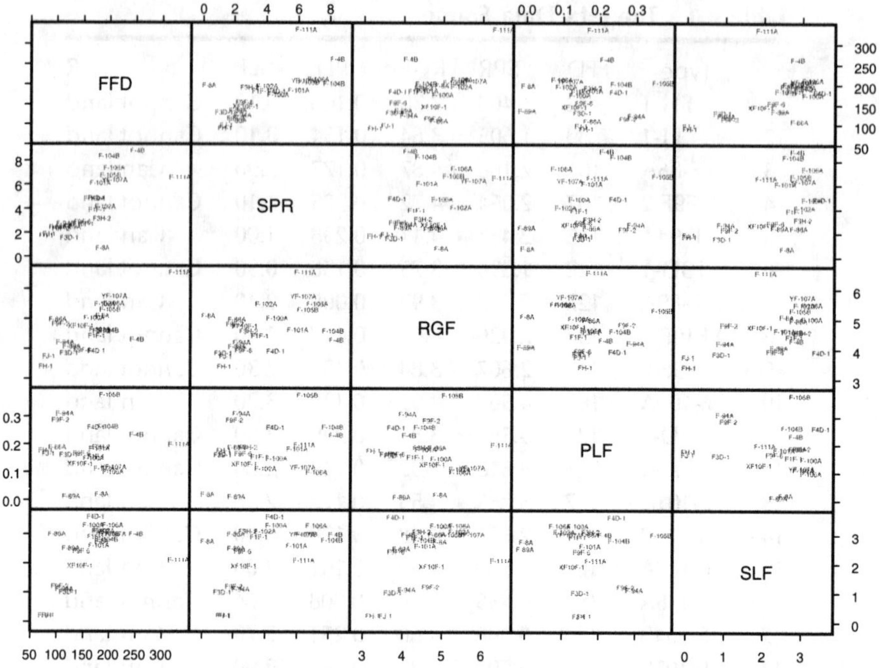

Figure 4.1   Scatterplot matrix of the variables in the jets data frame with points labelled by type of aircraft.

the plot with type of aircraft. With the command language, the required diagram can be found as follows:

```
attach(jets)
pairs(jets[,-c(1,7)],panel=function(x, y){
    points(x,y,pch=5)
    abline(lm(y~x))
    lines(lowess(x,y),lty=2)})
#
#first the Type and CAR variables are removed
#from the data frame prior to plotting
#In the panel function the points function is
#used to plot the data on each panel using
#plotting character number 5, the linear
#modelling function lm is used to regress the
#y variable on the x variable, and the result
#used by abline to plot the fitted regressions
```

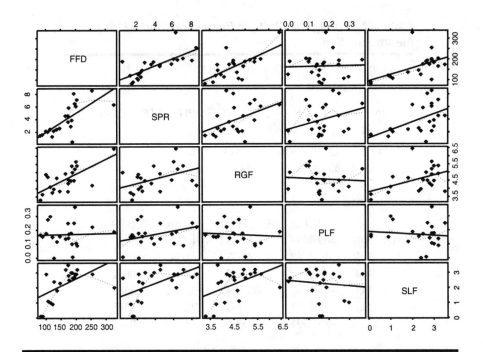

Figure 4.2   Scatterplot matrix of the variables in the jets data frame showing fitted lines and locally weighted regressions.

```
#finally lines and lowess are used to plot
#the results of a locally weighted regression
#with a different line type from the linear regression fit
```

The resulting diagram is shown in Figure 4.2.

In some of the panels the locally weighted regression fit differs considerably from the linear fit, confirming in some respect our suspicion about the form of the relationship between some pairs of variables. But for the moment, at least, we shall ignore this possible complication and find the results of the usual multiple regression model on the data. Using again the command line language, the model can be applied using the lm function.

```
jets.fit<-lm(FFD~SPR+RGF+PLF+SLF+CAR)
#stores the results of applying lm in jets.fit
summary(jets.fit)
```

The results are shown in Table 4.2. We see that the five explanatory variables account for about 80% (Multiple R-Squared is 0.7967) of the variability of FFD. The most predictive variables appear to be SPR, RGF, and CAR, although since the regression coefficients and their standard

Table 4.2 Fitting the Multiple Regression Model Using the lm Function

```
> jets.fit <- lm(FFD ~ SPR + RGF + PLF + SLF + CAR)
> summary(jets.fit)
```

Call: lm(formula = FFD ~ SPR + RGF + PLF + SLF + CAR)
Residuals:

Min	1Q	Median	3Q	Max
−44.48	−13.85	0.04965	14.96	64.91

Coefficients:

	Value	Std. Error	t value	Pr(>|t|)
(Intercept)	−89.4837	51.8632	−1.7254	0.1037
SPR	15.8788	3.9359	4.0343	0.0010
RGF	43.9766	11.2427	3.9116	0.0012
PLF	−84.1065	85.1941	−0.9872	0.3382
SLF	2.6411	8.3655	0.3157	0.7563
CAR	−20.9991	7.9932	−2.6271	0.0183

Residual standard error: 29.41 on 16 degrees of freedom
Multiple R-Squared: 0.7967
F-statistic: 12.54 on 5 and 16 degrees of freedom, the p-value is 0.00004569

Correlation of Coefficients:

	(Intercept)	SPR	RGF	PLF	SLF
SPR	0.1632				
RGF	−0.9042	−0.1540			
PLF	−0.3764	−0.4774	0.1328		
SLF	−0.0902	−0.4866	−0.2152	0.3039	
CAR	0.4949	−0.2474	−0.4388	0.0317	0.0157

errors are estimated *conditionally* on the other variables such an interpretation needs to be made with caution. Removing one of the nonsignificant explanatory variables and reestimating the remaining coefficients might lead to a different picture. We can investigate this further by fitting a variety of models and comparing their fits using the **anova** function; for example,

```
jets.fit1<-lm(FFD~SPR)
jets.fit2<-lm(FFD~SPR+RGF)
anova(jets.fit1,jets.fit2)
```

Table 4.3 Comparing Models Using the **anova** Function

```
> jets.fit1 <- lm(FFD ~ SPR)
> jets.fit2 <- lm(FFD ~ SPR + RGF)
> anova(jets.fit1, jets.fit2)
Analysis of Variance Table
```

Response: FFD

	Terms	Resid. Df	RSS	Test	Df	Sum of Sq	F Value	Pr(F)
1	SPR	20	32273.65					
2	SPR + RGF	19	20896.87	+RGF	1	11376.78	10.34407	0.004545782

Table 4.4 Comparing Models

```
> anova(jets.fit2, jets.fit3)
Analysis of Variance Table
```

Response: FFD

	Terms	Resid. Df	RSS	Test	Df	Sum of Sq	F Value
1	SPR + RGF	19	20896.87				
2	SPR + RGF + CAR	18	15043.81	+CAR	1	5853.064	7.003223
	Pr(F)						
1							
2	0.01641463						

gives the output shown in Table 4.3. The addition of RGF after SPR reduces the residual sum of squares by 11,377 which has an associated p value of 0.0045 — highly significant. Now let's see if adding CAR is worthwhile:

```
jets.fit3<-lm(FFD~SPR+RGF+CAR)
anova(jets.fit2,jets.fit3)
```

Here the results shown in Table 4.4 confirm that CAR adds significantly to the prediction of FFD after SPR and RGF have been used.

Repeating this process with the remaining two variables, PLF and SLF, shows that neither contribute significantly after SPR, RGF, and CAR. We can now find the estimated parameters and their standard errors in this reduced model from

```
summary(jets.fit3)
```

Table 4.5 Estimated Regression Coefficients etc. for Selected Model

> summary(jets.fit3)

Call: lm(formula = FFD ~ SPR + RGF + CAR)
Residuals:

Min	1Q	Median	3Q	Max
−58.58	−18.07	4.482	14.59	58.25

Coefficients:

	Value	Std. Error	t value	Pr(>|t|)
(Intercept)	−109.6063	47.2135	−2.3215	0.0322
SPR	14.9359	3.1038	4.8121	0.0001
RGF	47.4005	10.5445	4.4953	0.0003
CAR	−20.7819	7.8530	−2.6464	0.0164

Residual standard error: 28.91 on 18 degrees of freedom
Multiple R-Squared: 0.7791
F-statistic: 21.16 on 3 and 18 degrees of freedom, the p-value is 3.957e-006

Correlation of Coefficients:

	(Intercept)	SPR	RGF
SPR	−0.0099		
RGF	−0.9590	−0.2439	
CAR	0.5474	−0.2868	−0.4627

giving the results shown in Table 4.5. [Note that since S-PLUS codes the levels of CAR, 1 (can land) and −1 (cannot land) by default, the estimated parameter for CAR and its standard error need to be doubled to give the effect corresponding to a 1, 0 coding.]

The next stage in the analysis should be an examination of the residuals from fitting the chosen model to check on the normality and constant variance assumptions. The most useful plots of these residuals are as follows:

■ A plot of residuals against each explanatory variable in the model. The presence of a nonlinear relationship, for example, may suggest that a higher-order term in the explanatory variable should be considered.

■ A plot of residual against fitted values. If the variance of the residuals appears to increase with predicted value, a transformation of the response variable may be in order.

■ A normal probability plot of the residuals. After all the systematic variation has been removed from the data, the residuals should look like a sample from a standard normal distribution. A plot of the ordered residuals against the expected order statistics from a normal distribution provides a graphical check of this assumption.

Unfortunately, the basic 'observed-fitted' residuals suffer from having a distribution that is scale dependent because their variance is a function of both σ^2 and the diagonal values of the so-called hat matrix, $\mathbf{H}$, given by

$$\mathbf{H} = \mathbf{X}(\mathbf{X'X})^{-1}\mathbf{X'} \qquad (4.3)$$

where $\mathbf{X}$ is the matrix introduced in Section 4.2. (See Cook and Weisberg, 1982, for a full explanation of the hat matrix.) Consequently, it is rather more useful to work with standardized residuals $r_i^{(std)}$ given by

$$r_i^{(std)} = \frac{y_i - \hat{y}_i}{s\sqrt{1 - h_{ii}}} \qquad (4.4)$$

where s^2 is the usual estimate of σ^2 obtained as the residual mean square from the model fitting, y_i is the observed response value, $\hat{y}_i$ is the value predicted from the fitted model, and h_{ii} is the ith element in the main diagonal of $\mathbf{H}$. These standardised residuals are easily calculated using the **lm.influence** function:

```
s<-summary(jets.fit3)$sigma
#gives the required value of s in 4.4
h<-lm.influence(jets.fit3)$hat
#gives the vector of the main diagonal elements
#of the hat matrix in 4.3
jets.res<-residuals(jets.fit3)/(s*sqrt(1-h))
#the residuals function gives the basic observed fitted
#residuals from jets.fit3
```

We may now plot these standardised residuals in a number of ways:

```
#construct plots of the standardised residuals
#against each of the three explanatory variables
par(mfrow=c(1,3))
#set up plotting area to take all three plots
```

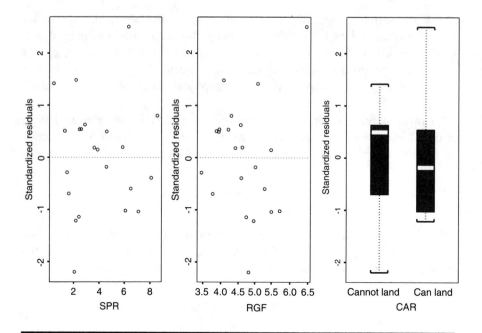

Figure 4.3 Standardised residuals for regression model fitted to jet fighter data plotted against variables SPR, RGF, and CAR.

```
plot(SPR,jets.res,ylab="Standardised residuals")
abline(h=0,lty=2)
#add dotted horizontal line at y=0
plot(RGF,jets.res,ylab="Standardised residuals")
abline(h=0,lty=2)
plot(CAR,jets.res,ylab="Standardised residuals")
```

This leads to Figure 4.3. It is also helpful to have the numerical values of the observed values, fitted values, and the standardised residuals; these are all given in Table 4.6 and were obtained using

```
cbind(FFD,predict(jets.fit3),jets.res)
#the predict function gives the fitted value from the
#model; the function cbind combines the three
#vectors into a matrix
```

It is standardised residuals outside of the interval (–2, 2) that give most cause for concern. Here there are two, aircraft number 4 (F9F-2), where the predicted value of FFD is considerably higher than the observed value,

Table 4.6 Observed and Predicted Values of FFD and the Corresponding Residual Values

```
> cbind(FFD, predict(jets.fit3), jets.res)
```

	FFD		jets.res
1	82	89.5231	-0.2887851
2	89	107.6855	-0.6938394
3	101	132.8333	-1.2158725
4	107	165.5842	-2.1976956
5	115	101.2748	0.5402804
6	122	108.2545	0.5095121
7	127	90.3969	1.4806494
8	137	167.8224	-1.1429785
9	147	132.1314	0.5431492
10	166	171.0346	-0.1829026
11	174	160.7714	0.4965961
12	175	169.9838	0.1827579
13	177	171.7859	0.1940028
14	184	166.8140	0.6282893
15	187	183.0519	0.1468159
16	189	154.4998	1.4155201
17	194	203.7158	-0.3911602
18	197	213.2078	-0.5962569
19	201	228.2499	-1.0206190
20	204	231.6942	-1.0368842
21	255	237.9299	0.8020890
22	328	269.7550	2.5074932

and aircraft number 22 (F-111A), where the reverse is the case. Apart from these two outliers, the residuals in Figure 4.3 look reasonably well behaved.

Finally, we can look at a normal probability plot of the standardised residuals using

```
qqnorm(jets.res)
qqline(jets.res)
```

leading to Figure 4.4.

The interpretation of a normal probability plot is often unclear, particularly when, as with these data on jet fighters, there are only a relatively small number of observations. The plot can be made more informative by supplementing it with a confidence interval, as suggested in Atkinson

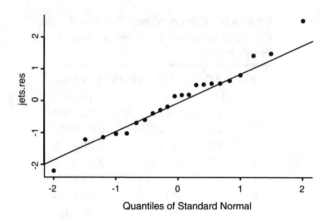

Figure 4.4 Normal probability plot of standardised residuals from model fitted to jet fighter data.

(1987) and Cook and Weisberg (1982). In essence, the procedure involves the generation of *m pseudo-residual* vectors, **e,** from

$$e_k = \left(I - H\right)\epsilon_k, \quad k = 1,\ldots,m \tag{4.5}$$

where the ϵ_k are $n \times 1$ vectors of standard normal variates.

Suitable code for the construction of such an enhanced probability plot using 100 simulations is as follows:

```
#first change strings for categories of CAR to 0 and 1
#
jets$CAR<-factor(jets$CAR,levels=c("Can not land","Can land"),
labels=c(0,1))
#
#
jets.one<-cbind(rep(1,22),jets[,c(3,4,7)])
#store the three explanatory variables SPR, RGF and CAR
#chosen for the final model along with
#a column of ones to get the usual X matrix used in
#multiple regression
#
#
jets.one<-as.matrix(jets.one)
#
```

```
#get hat matrix as defined in Equation 4.3
#
jets.hat<-
jets.one%*%solve(t(jets.one)%*%jets.one)%*%t(jets.one)
#
#construct a diagonal matrix as required in Equation 4.5
#
ident<-diag(22)
#
#generate 100 sets of 22 standard normal variables, i.e., 2200
normal variates
#
set.seed(547)
epsilon<-matrix(rnorm(100*22,0,1),ncol=100)
#
#apply Equation 4.5 to get pseudo-residuals
#
e<-(ident-jets.hat)%*%epsilon
#
#now tranpose e and sort the elements in the columns
#save the 5th and 95th ranked values and then get the
#range of each column to give the required values for the
#95% CI for the probability plot
#
e<-t(e)
e<-apply(e,2,sort)
e<-e[5:95,]
E<-apply(e,2,range)
#
#now get a normal probability plot of the standardised residuals
#with the confidence limits
#
win.graph()
ylim<-range(jets.res,E[1,],E[2,])
qqnorm(jets.res,ylim=ylim,ylab="Standardised residuals")
par(new=T)
qqnorm(sort(E[1,]),type="l",axes=F,ylim=ylim,xlab=" ",ylab=" ")
par(new=T)
qqnorm(sort(E[2,]),type="l",axes=F,ylim=ylim,xlab=" ",ylab=" ")
```

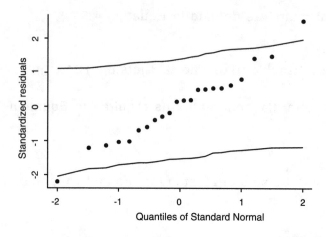

Figure 4.5   Normal probability plot of standardised residuals from model fitted to jet fighter data with added confidence interval.

The resulting plot is shown in Figure 4.5. The first and last observations fall outside the constructed confidence interval, and it may be worthwhile rerunning the regression analysis after removing these two observations.

Exercises

4.1. Investigate the use of the **plot** function with a regression object from using the **lm** function. Try, for example, both **plot(jets.fit3)** and **plot(jets.fit3,ask=T)**.

4.2. Investigate the use of the **Stepwise Linear Regression** dialog for selecting a parsimonious model with which to describe the jet fighter data.

4.3. Other researchers dealing with the jets data have used a log transformation of FFD. Examine whether this approach leads to conclusions different from those reached in the text.

4.4. Many other regression diagnostics are available apart from those described in the text; see Rawlings (1988) for a good description. Use the **Linear Regression** dialog to investigate the use of some of these data, particularly *Cook's distance* and *partial residuals*, on the final model selected in the text for the jet fighter.

Chapter 5

Logistic Regression: Psychiatric Caseness and Mortgage Default

5.1 Description of Data

In this chapter we shall look at two data sets from very different application areas. The first, shown in Table 5.1, consists of data collected during a study of a psychiatric screening questionnaire, the *General Health Questionnaire* or GHQ (see Goldberg, 1972), designed to help identify possible psychiatric 'caseness'.

The second data set, kindly supplied by Professor David Hand and shown in Table 5.2, arises from an investigation into the reasons for mortgage default.

In the first of the two data sets, interest lies in assessing whether GHQ score is predictive of 'caseness' and whether the sex of a subject plays a role in this prediction. In the second data set the main question is whether any of the four explanatory variables might be used to identify mortgage loans at risk of default.

5.2 Logistic Regression and Generalised Linear Models

In essence, the same general question is addressed by both data sets introduced in the previous section, namely, 'can a particular response

Table 5.1 Psychiatric Data

ghqscore	sex	case	nocase
0	0	4	80
1	0	4	29
2	0	8	15
3	0	6	3
4	0	4	2
5	0	6	1
6	0	3	1
7	0	2	0
8	0	3	0
9	0	2	0
10	0	1	0
0	1	1	36
1	1	2	25
2	1	2	8
3	1	1	4
4	1	3	1
5	1	3	1
6	1	2	1
7	1	4	2
8	1	3	1
9	1	2	0
10	1	2	0

Gender is coded 0 for women and 1 for men.

variable be predicted from a set of explanatory variables?' This is the same question that is addressed by the multiple regression model that was the subject of the previous chapter, so readers might ask why are we considering it again here? The reason is that the response variable for each data set in Section 5.1 is now binary rather than continuous — case or not case for the data in Table 5.1 and defaulted on loan or not for the second. We could, of course, simply ignore this aspect of the data and, as in the previous chapter, model the expected value of the response as a linear function of the explanatory variables. For a binary variable this expected value is simply the probability of the 'one' category of the variable, so for the psychiatric data, for example, our model would be

$$p = \Pr(\text{of being a case}) = \beta_0 + \beta_1 \text{sex} + \beta_2 \text{ghq} \qquad (5.1)$$

Table 5.2 Mortgage Default Data

	EVER90	FICO	LTV	INCOME	CA
1	1	576	0.86	5	0
2	1	678	0.90	3	1
3	1	693	0.80	5	0
4	1	669	0.75	4	0
5	1	542	0.95	4	0
6	1	566	0.95	3	0
7	1	643	0.84	5	0
8	0	785	0.58	5	0
9	1	461	0.80	4	0
10	0	785	0.85	5	0
11	1	620	0.95	3	0
12	0	792	0.79	3	0
13	0	748	0.90	4	0
14	1	661	0.75	5	0
15	0	753	0.70	5	0
16	0	747	0.95	4	0
17	0	720	0.88	5	0
18	0	717	0.89	4	0
19	0	763	0.90	5	0
20	1	782	0.90	4	1
21	0	758	0.42	5	0
22	0	697	0.80	5	0
23	1	690	0.75	2	0
24	0	746	0.90	5	1
25	0	679	0.70	5	0
26	0	767	0.80	2	0
27	0	733	0.88	5	0
28	0	679	0.75	2	0
29	0	765	0.36	3	0
30	0	678	0.95	5	0
31	0	782	0.62	4	0
32	0	777	0.80	5	0
33	0	765	0.63	5	0
34	0	769	0.59	5	0
35	0	720	0.38	5	0
36	0	758	0.90	4	0
37	0	724	0.70	5	0
38	1	611	0.90	5	0
39	0	639	0.61	5	0
40	0	761	0.50	3	0

Table 5.2 (Continued)   Mortgage Default Data

	EVER90	FICO	LTV	INCOME	CA
41	1	549	0.66	3	0
42	0	639	0.68	6	0
43	0	719	0.61	5	0
44	0	801	0.73	2	0
45	0	765	0.90	5	0
46	1	598	0.95	5	0
47	1	603	0.95	4	0
48	0	710	0.67	2	0
49	0	774	0.48	5	0
50	0	738	0.73	5	0
51	1	676	0.73	3	0
52	0	768	0.78	5	0
53	0	765	0.90	5	0
54	0	766	0.87	5	0
55	0	729	0.90	4	0
56	0	766	0.68	5	0
57	1	671	0.75	1	0
58	0	782	0.80	2	0
59	1	756	0.75	2	0
60	0	791	0.89	4	0
61	1	650	0.85	3	0
62	0	704	0.75	3	0
63	1	664	0.80	8	1
64	0	762	0.62	4	0
65	0	733	0.70	3	0
66	1	634	0.95	5	1
67	0	797	0.26	3	1
68	1	624	0.90	5	1
69	0	749	0.70	2	0
70	0	687	0.95	3	0
71	0	732	0.72	4	1
72	0	669	0.80	5	0
73	0	692	0.79	5	0
74	0	777	0.80	3	1
75	0	682	0.47	4	1
76	0	772	0.80	5	0
77	0	809	0.75	5	0
78	0	728	0.74	5	0
79	0	669	0.67	5	1
80	1	692	0.75	4	0

Table 5.2 (Continued) Mortgage Default Data

	EVER90	FICO	LTV	INCOME	CA
81	0	721	0.75	5	0
82	1	742	0.69	5	0
83	0	687	0.79	2	0
84	0	696	0.83	5	1
85	1	586	0.90	5	0
86	0	727	0.80	3	1
87	0	725	0.80	7	0
88	1	652	0.95	5	1
89	1	722	0.80	5	1
90	1	606	0.80	4	0
91	0	761	0.80	5	0
92	0	777	0.80	5	0
93	0	629	0.29	5	0
94	1	742	0.74	3	0
95	0	752	0.61	2	0

EVER90: takes value 1 if loan has ever been 90 days past
   due date during a specified time period, 0 otherwise.
FICO: a credit score provided by Fair Isaac Company.
LTV: ratio of loan amount to value of home.
INCOME: income category of borrower.
CA takes value 1 if home in California, 0 otherwise.

We could then estimate the three regression coefficients, β_0, β_1, and β_2 by least squares as in the previous chapter. Unfortunately, this approach is not satisfactory for two reasons;

1. It can lead to fitted values for the probability of being a case outside the interval (0,1).
2. The normality assumption needed in Chapter 4 to derive significance tests for model parameters is clearly not reasonable for a binary response.

A more suitable approach to modelling a binary response is to use *logistic regression*. Here the expected value of the response variable is not modelled directly; instead, the *logistic function* of the expected value is used, so the model for the psychiatric data becomes

$$\log \frac{p}{1-p} = \beta_0 + \beta_1 \text{sex} + \beta_2 \text{ghq} \tag{5.2}$$

(β_0, β_1, and β_2 will not, of course, be the same as in Equation 5.1).

The logistic transformation gives the *log odds* of being a case, and by now assuming a *binomial distribution* for the observed response, the parameters in Equation 5.2 can be estimated by maximum likelihood (see Collett, 1991, for details). In terms of p, the model in Equation 5.2 can be rewritten as

$$p = \frac{\exp\left[\beta_0 + \beta_1 \text{sex} + \beta_2 \text{ghq}\right]}{1 + \exp\left[\beta_0 + \beta_1 \text{sex} + \beta_2 \text{ghq}\right]} \tag{5.3}$$

Fitted values of p will now lie in the interval $(0,1)$ as required.

The estimated regression coefficients in a logistic regression model give the estimated change in the log odds corresponding to a unit change in the corresponding explanatory variable. For easier interpretation, the parameters are usually exponentiated to give the estimated change in the odds, conditional on the other variables remaining constant.

Logistic regression and the other modelling procedures used in earlier chapters, analysis of variance and multiple regression, can all be shown to be special cases of the *generalised linear model* formulation described in detail in McCullagh and Nelder (1989). In essence, this approach postulates a linear model for a suitable transformation of the expected value of a response variable and allows for a variety of different error distributions. The possible transformations are known as *link functions*. For multiple regression and analysis of variance, for example, the link function is simply the identity function, so the expected value is modelled directly, and the corresponding error distribution is normal. For logistic regression, the link function is logistic and the appropriate error distribution is the binomial. Many other possibilities are opened up by the generalised linear model formulation; see McCullagh and Nelder (1989) for full details.

5.3 Analysis Using S-PLUS

Logistic regression is available in S-PLUS via the generalised linear model function, **glm**, or by using the **Logistic Regression** dialog. Both possibilities will be used in the following subsections.

5.3.1 GHQ Data

We shall assume that the GHQ data are available as an S-PLUS data frame **ghq**, the contents of which are shown in Table 5.3. (Note that the data have been grouped to give the total number of cases and noncases for

Table 5.3   Contents of ghq Data Frame

> *ghq*

	ghqscore	sex	case	nocase
1	0	0	4	80
2	1	0	4	29
3	2	0	8	15
4	3	0	6	3
5	4	0	4	2
6	5	0	6	1
7	6	0	3	1
8	7	0	2	0
9	8	0	3	0
10	9	0	2	0
11	10	0	1	0
12	0	1	1	36
13	1	1	2	25
14	2	1	2	8
15	3	1	1	4
16	4	1	3	1
17	5	1	3	1
18	6	1	2	1
19	7	1	4	2
20	8	1	3	1
21	9	1	2	0
22	10	1	2	0

each combination of the two explanatory variables; originally, however, the data would have consisted of whether or not each subject was a case.)

We shall use the command language approach to analyse these data using the **glm** function to apply logistic regression. To begin, we shall consider only the single explanatory variable **ghqscore**.

```
p<-case/(case+nocase)
ghq<-data.frame(ghq,p)
attach(ghq)
#first calculate the proportion of cases and then
#insert into the ghq data frame
fit.lin<-glm(p~ghqscore)
#the default when using glm is the equivalent of
#using the lm function, i.e., simple linear or
#multiple regression. We fit this model first
```

```
#to be able to compare the results with those
#obtained from logistic regression using
fit.log<-glm(p~ghqscore,family=binomial,
weights=case+ncase)
#the argument family=binomial implements a
# binomial distribution and if no link is
#specified uses the logistic with the binomial choice.
#Since the observations are grouped the weights
#argument must be used to give the total number
#of observations on which each observed proportion is
#based.
summary(fit.lin)
summary(fit.log)
```

The results are shown in Table 5.4. In both models the regression coefficient for the GHQ variable is highly significant. The *deviance* term that occurs in Table 5.4 measure lack of fit. Specifically, it is minus twice the difference between the maximized log-likelihood of the model and the maximum likelihood achievable, i.e., the maximized likelihood of the full or saturated model. For normal distributions, the deviance is simply the well-known *residual sum of squares*. The difference in deviance between two competing models can be used to compare them, by referring its value to a chi-square distribution with degrees of freedom equal to the difference in the degrees of freedom of the two models.

Now we can use the **predict** function to look at the predicted values from each model:

```
lin.pred<-predict(fit.lin)
log.pred<-exp(predict(fit.log))/(1+exp(predict(fit.log)))
cbind(lin.pred,log.pred)
```

The fitted values are shown in Table 5.5. (Note that the first 11 equal the second 11 since they correspond to the same GHQ scores, 0 to 10.) A problem that immediately becomes apparent with the simple linear regression model is that some of its predictions are greater than 1. A graphical comparison of the fits along with the observed proportions is also helpful here, and can be obtained as follows:

```
Case<-case[1:11]+case[12:22]
Nocase<-nocase[1:11]+nocase[12:22]
P<-Case/(Case+Nocase)
#calculate the observed proportions of cases
```

Table 5.4   Linear and Logistic Regression Models for the Psychiatric Data

```
> fit.lin <- glm(p ~ ghqscore)
> fit.log <- glm(p ~ ghqscore, family = binomial, weights = case + nocase)
> summary(fit.lin)
```

Call: glm(formula = p ~ ghqscore)
Deviance Residuals:

Min	1Q	Median	3Q	Max
–0.2150518	–0.1162369	–0.03278505	0.1217964	0.2516149

Coefficients:

	Value	Std. Error	t value
(Intercept)	0.1143407	0.05922838	1.930505
ghqscore	0.1002370	0.01001142	10.012266

(Dispersion Parameter for Gaussian family taken to be 0.0220503)
    Null Deviance: 2.651447 on 21 degrees of freedom
Residual Deviance: 0.4410058 on 20 degrees of freedom
Number of Fisher Scoring Iterations: 1

Correlation of Coefficients:

	(Intercept)
ghqscore	–0.8451543

```
> summary(fit.log)
```

Call: glm(formula = p ~ ghqscore, family = binomial, weights = case + nocase)
Deviance Residuals:

Min	1Q	Median	3Q	Max
–1.768974	–0.7230615	0.1255168	0.5306368	1.757787

Coefficients:

	Value	Std. Error	t value
(Intercept)	–2.7107322	0.27243281	–9.950094
ghqscore	0.7360353	0.09456846	7.783094

(Dispersion Parameter for Binomial family taken to be 1)
    Null Deviance: 130.3059 on 21 degrees of freedom
Residual Deviance: 16.23682 on 20 degrees of freedom
Number of Fisher Scoring Iterations: 5

Correlation of Coefficients:

	(Intercept)
ghqscore	–0.7324319

Table 5.5   Predicted Values for Linear and Logistic Regression Models Fitted to the Psychiatric Data

```
> lin.pred <- predict(fit.lin)
> log.pred <- exp(predict(fit.log))/(1 + exp(predict(fit.log)))
> cbind(lin.pred, log.pred)
```

	lin.pred	log.pred
1	0.1143407	0.06234304
2	0.2145777	0.12188529
3	0.3148147	0.22466904
4	0.4150518	0.37692367
5	0.5152888	0.55808875
6	0.6155258	0.72500870
7	0.7157628	0.84624903
8	0.8159999	0.91993872
9	0.9162369	0.95998065
10	1.0164739	0.98042217
11	1.1167110	0.99052540
12	0.1143407	0.06234304
13	0.2145777	0.12188529
14	0.3148147	0.22466904
15	0.4150518	0.37692367
16	0.5152888	0.55808875
17	0.6155258	0.72500870
18	0.7157628	0.84624903
19	0.8159999	0.91993872
20	0.9162369	0.95998065
21	1.0164739	0.98042217
22	1.1167110	0.99052540

```
#corresponding to each GHQ score, after
#combining over males and females
par(pty="s")
ylim<-range(lin.pred,log.pred,P)
plot(ghqscore[1:11],lin.pred[1:11],xlab="GHQ",
ylab="Probability of case",type="l",ylim=ylim)
#plots ghqscore against the predictions made
#by the simple linear regression
#model and draws a line through the points.
lines(ghqscore[1:11],log.pred[1:11],lty=2)
#adds a dotted line for the logistic regression
```

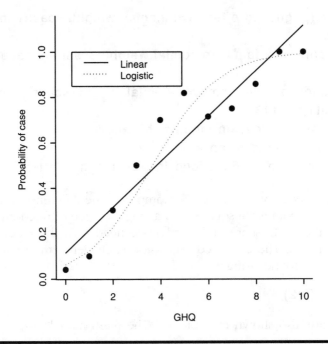

Figure 5.1 GHQ data showing observations and fitted linear and locally weighted regressions.

```
#predictions
points(ghqscore[1:11],P)
#adds the observed proportions to the plot
legend(0.3,1,c("Linear","Logistic"),lty=1:2)
#adds a legend at a convenient place on the plot.
```

The resulting graph is shown in Figure 5.1. We now see that in addition to giving predicted values greater than 1, the linear model provides a very poor description of the observed proportions.

Now let's consider sex as well as the GHQ score and compare the fit of the following three models.

1. GHQ score
2. GHQ score + sex
3. GHQ score + sex + sex × GHQ score

We now fit each of these three models using the **glm** function and then compare their fit using the **anova** function.

```
fit1<-glm(p~ghqscore,family=binomial,weights=case+nocase)
fit2<-glm
p~ghqscore+sex,family=binomial,weights=case+nocase)
fit3<-
glm(p~ghqscore*sex,family=binomial,weights=case+nocase)
anova(fit1,fit2,fit3)
#for logistic regression models the anova
#function will lead to an analysis of
#deviance table as described in McCullagh and Nelder.
```

The results are shown in Table 5.6. Comparing the difference in deviances of the models with a chi-square with a single degree of freedom indicates that both the GHQ score and gender are needed in the model but not their interaction. The estimated regression coefficients and their standard errors can be found using

```
summary(fit2)
```

and they are also shown in Table 5.6. The fitted models are

$$\log\left(\text{odds of being a case}\right) = -2.49 + 0.78 \text{ ghqscore} - 0.94 \text{ sex} \quad (5.4)$$

The standard error of the regression coefficient for the GHQ score is 0.099, so an approximate 95% confidence interval for the coefficient is (0.58, 0.98). An increase of 1 in GHQ score raises the log odds of being a case by between 0.58 and 0.98, conditional on gender. This becomes easier to interpret after the two limits are exponentiated so that we can relate the change to odds rather than log (odds):

```
exp(c(0.58,0.98))
```

gives the interval (1.79, 2.66). So a 1-unit increase in GHQ score increases the odds of being a case by about twofold. The same series of calculations for gender leads to the confidence interval for the odds ratio of being a case in males and females of (0.25, 0.60), conditional on the GHQ score. Clearly, for a given GHQ score, men are less likely to be cases than women.

As with the multiple regression model considered in the previous chapter, fitting a logistic regression model is not complete without checking on model assumptions by examining the properties of some suitably defined residuals. There are a number of ways in which a fitted logistic model may be inadequate:

Table 5.6 Three Logistic Models for the Psychiatric Data Compared

```
> fit1 <- glm(p ~ ghqscore, family = binomial, weights = case +
nocase, data = ghq)
> fit2 <- glm(p ~ ghqscore + sex, family = binomial, weights = case +
      nocase, data = ghq)
> fit3 <- glm(p ~ ghqscore * sex, family = binomial, weights = case +
      nocase, data = ghq)
> anova(fit1, fit2, fit3)
Analysis of Deviance Table
```

Response: p

	Terms	Resid. Df	Resid. Dev	Test	Df	Deviance
1	ghqscore	20	16.2			
2	ghqscore + sex	19	11.1	+sex	1	5.12
3	ghqscore * sex	18	8.8	+ghqscore:sex	1	2.35

```
> summary(fit2)
```

Call: glm(formula = p ~ ghqscore + sex, family = binomial, data = ghq, weights = case + nocase)
Deviance Residuals:

Min	1Q	Median	3Q	Max
−1.4	−0.394	0.188	0.432	1.33

Coefficients:

	Value	Std. Error	t value
(Intercept)	−2.494	0.282	−8.85
ghqscore	0.779	0.099	7.87
sex	−0.936	0.434	−2.16

(Dispersion Parameter for Binomial family taken to be 1)

Null Deviance: 130 on 21 degrees of freedom

Residual Deviance: 11.1 on 19 degrees of freedom

Number of Fisher Scoring Iterations: 5

Correlation of Coefficients:

	(Intercept)	ghqscore
ghqscore	−0.611	
sex	−0.220	−0.305

■ The linear function of the explanatory variables may be inadequate; for example, one or more of the explanatory variables may need to be transformed.

■ The logistic transformation of the response probability may not be appropriate.

■ The data may contain outliers that are not well fitted by the model.

■ The assumption of a binomial distribution may not be correct.

In logistic regression there are several types of residuals that may be useful for assessing one or other of those potential difficulties. Here, we shall define two and then illustrate their use on the GHQ data. We assume there are n observations of the form y_i/n_i, $i = 1, 2, \ldots, n$ and that the corresponding fitted value of y_i is $\hat{y}_i = n_i\hat{p}_i$. The ith row residual is then the difference $y_i - \hat{y}_i$ and provides information about how well the model fits each particular observation. But because they are based on different n_is and for other reasons explained in Collett (1991), the raw residuals are difficult to interpret. Better are the *Pearson residuals* and the *deviance residuals*:

■ Pearson residuals

$$\frac{y_i - n_i\hat{p}_i}{\sqrt{n_i\hat{p}_i(1 - \hat{p}_i)}} \tag{5.5}$$

■ Deviance residuals

$$\text{sign}(y_i - \hat{y}_i)\left[2y_i\log\left(\frac{y_i}{\hat{y}_i}\right) + 2(n - y_i)\log\left(\frac{n - y_i}{n - \hat{y}_i}\right)\right]^{1/2} \tag{5.6}$$

where $\text{sign}(y_i - \hat{y}_i)$ is the function that makes d_i positive when $(y_i \geq \hat{y}_i)$ and negative when $(y_i > \hat{y}_i)$.

Both residuals are best used after standardising by dividing by $\sqrt{1 - h_i}$ where h_i is the ith diagonal element of the 'hat matrix' in logistic regression (see Collett, 1991).

There are various ways of plotting the residuals that give different insights into possible model inadequacies. Three possibilities are as follows:

■ *Index plot*: plot of residuals against observation number or index. Useful for the detection of outliers.

■ Plot of residuals against values of the linear predictor. The occurrence of a systematic pattern in the plot suggests the model is incorrect in some way.

■ Plot of residuals against explanatory variables in the model may help to identify whether the variable needs to be transformed.

Suitable commands for constructing each of these plots for the deviance residuals from the logistic regression model with the explanatory variables **ghqscore and sex** are as follows

```
resid<-residuals(fit2,type="deviance")
#store deviance residuals from the
#fitted model in resid
resides<-resid/sqrt(1-lm.influence(fit2)$hat)
#uses lm.influence to get the diagonal
#elements of the logistic regression
#equivalent of the hat matrix introduced
#in Chapter 4
par(mfrow=c(1, 3))
#set up plotting area to take three plots
#side-by-side
plot(1:22,resides,xlab="Observation No.",
ylab="Standardised deviance residual")
plot(predict(fit2),resides,xlab="Predicted value",
ylab="Standardised deviance residual")
plot(ghqscore,resides,xlab="GHQ",ylab=
"Standardised deviance residual")
```

The resulting diagram is shown in Figure 5.2. There is no discernable pattern in these residuals that would give cause for concern.

5.3.2 Mortgage Default Data

We assume that the mortgage default data are available as the data frame **credit**, the contents of which are shown in Table 5.7. (The information about the income categories in the **INCOME** variable is confidential, although it is known that the category labels are monotone with increasing income. Here we shall simply use these labels as the values of a continuous variable in the modelling process.)

We shall now use the **Logistic Regression** dialog to undertake the logistic regression. To access this dialog:

- ■ Click on **Statistics**.
- ■ Select **Regression**.
- ■ Select **Logistic**.

In the dialog choose the **credit** data frame. As the dependent variable, select **EVER90**, and as independent variables, highlight **FICO, LTV, INCOME**, and **CA**. The dialog box now appears as shown in Figure 5.3.

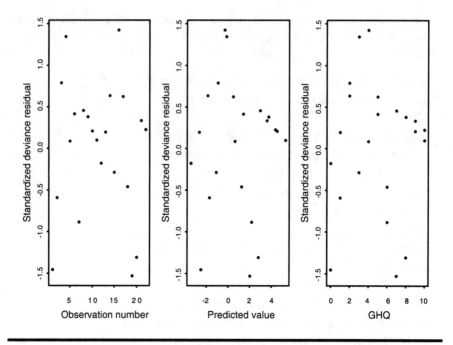

Figure 5.2 Residual plots for logistic regression model including ghqscore **and sex fitted to GHQ data**

Click **OK** to obtain the results shown in Table 5.8. It appears that the variables FICO and LTV are of most importance in predicting mortgage default. Largely as an exercise in using some other S-PLUS features, we will now demonstrate how to construct some informative graphics for displaying the relationships involved.

To display graphically the relationship between the variables FFCO and LTV and the probability of mortgage default, we first need to use the **Logistic Regression** dialog again to fit a model using only these two variables, and to save the fitted values.

- Click on **Statistics**.
- Select **Regression**.
- Select **Logistic**.
- Select **credit** data frame.
- Select **EVER90** as dependent variable.
- Select **FICO** and **LTV** as independent variables.
- Click on **Results** tab.
- In **Saved Results** section choose **Fitted Values**, and in **Save In** slot select **credit** data frame.
- Click **OK**.

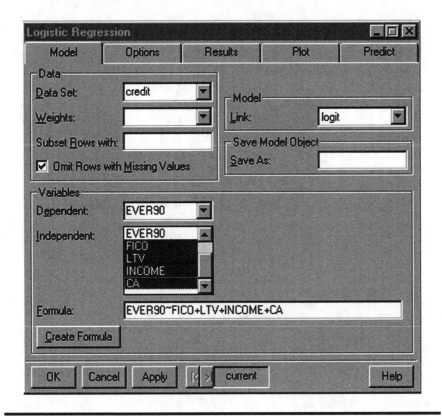

Figure 5.3 Logistic Regression dialog showing a model for the mortgage default data.

The results will appear in a Report file and the new **credit** data frame, which now has the fitted values incorporated, is also shown:

- Highlight **FICO**.
- **Ctrl click** on **LTV**.
- **Ctrl click** on **fit**.
- Click on

to access **3D graph** palette.
- Select **Drop Line Scatter** as shown in Figure 5.4.

The results shown in Figure 5.5, which shows the relationship between the probability of mortgage default and FICO and LTV. A somewhat clearer

Table 5.7   Contents of credit Data Frame

> *credit*

	row.labels	EVER90	FICO	LTV	INCOME	CA
1	1	1	576	0.86	5	Other
2	2	1	678	0.90	3	California
3	3	1	693	0.80	5	Other
4	4	1	669	0.75	4	Other
5	5	1	542	0.95	4	Other
6	6	1	566	0.95	3	Other
7	7	1	643	0.84	5	Other
8	8	0	785	0.58	5	Other
9	9	1	461	0.80	4	Other
10	10	0	785	0.85	5	Other
11	11	1	620	0.95	3	Other
12	12	0	792	0.79	3	Other
13	13	0	748	0.90	4	Other
14	14	1	661	0.75	5	Other
15	15	0	753	0.70	5	Other
16	16	0	747	0.95	4	Other
17	17	0	720	0.88	5	Other
18	18	0	717	0.89	4	Other
19	19	0	763	0.90	5	Other
20	20	1	782	0.90	4	California
21	21	0	758	0.42	5	Other
22	22	0	697	0.80	5	Other
23	23	1	690	0.75	2	Other
24	24	0	746	0.90	5	California
25	25	0	679	0.70	5	Other
26	26	0	767	0.80	2	Other
27	27	0	733	0.88	5	Other
28	28	0	679	0.75	2	Other
29	29	0	765	0.36	3	Other
30	30	0	678	0.95	5	Other
31	31	0	782	0.62	4	Other
32	32	0	777	0.80	5	Other
33	33	0	765	0.63	5	Other
34	34	0	769	0.59	5	Other
35	35	0	720	0.38	5	Other
36	36	0	758	0.90	4	Other
37	37	0	724	0.70	5	Other
38	38	1	611	0.90	5	Other
39	39	0	639	0.61	5	Other

Table 5.7 (Continued)   Contents of credit Data Frame

> *credit*

	row.labels	EVER90	FICO	LTV	INCOME	CA
40	40	0	761	0.50	3	Other
41	41	1	549	0.66	3	Other
42	42	0	639	0.68	6	Other
43	43	0	719	0.61	5	Other
44	44	0	801	0.73	2	Other
45	45	0	765	0.90	5	Other
46	46	1	598	0.95	5	Other
47	47	1	603	0.95	4	Other
48	48	0	710	0.67	2	Other
49	49	0	774	0.48	5	Other
50	50	0	738	0.73	5	Other
51	51	1	676	0.73	3	Other
52	52	0	768	0.78	5	Other
53	53	0	765	0.90	5	Other
54	54	0	766	0.87	5	Other
55	55	0	729	0.90	4	Other
56	56	0	766	0.68	5	Other
57	57	1	671	0.75	1	Other
58	58	0	782	0.80	2	Other
59	59	1	756	0.75	2	Other
60	60	0	791	0.89	4	Other
61	61	1	650	0.85	3	Other
62	62	0	704	0.75	3	Other
63	63	1	664	0.80	8	California
64	64	0	762	0.62	4	Other
65	65	0	733	0.70	3	Other
66	66	1	634	0.95	5	California
67	67	0	797	0.26	3	California
68	68	1	624	0.90	5	California
69	69	0	749	0.70	2	Other
70	70	0	687	0.95	3	Other
71	71	0	732	0.72	4	California
72	72	0	669	0.80	5	Other
73	73	0	692	0.79	5	Other
74	74	0	777	0.80	3	California
75	75	0	682	0.47	4	California
76	76	0	772	0.80	5	Other
77	77	0	809	0.75	5	Other
78	78	0	728	0.74	5	Other

Table 5.7 (Continued) Contents of credit Data Frame

> *credit*

	row.labels	EVER90	FICO	LTV	INCOME	CA
79	79	0	669	0.67	5	California
80	80	1	692	0.75	4	Other
81	81	0	721	0.75	5	Other
82	82	1	742	0.69	5	Other
83	83	0	687	0.79	2	Other
84	84	0	696	0.83	5	California
85	85	1	586	0.90	5	Other
86	86	0	727	0.80	3	California
87	87	0	725	0.80	7	Other
88	88	1	652	0.95	5	California
89	89	1	722	0.80	5	California
90	90	1	606	0.80	4	Other
91	91	0	761	0.80	5	Other
92	92	0	777	0.80	5	Other
93	93	0	629	0.29	5	Other
94	94	1	742	0.74	3	Other
95	95	0	752	0.61	2	Other

picture can be obtained by selecting **Spline Surface** from the **3D** graphics palette (Figure 5.6). Here the resulting diagram is shown in Figure 5.7. We see that as FICO and LTV increase, the probability of mortgage default decreases.

Table 5.8 Logistic Regression Model for Mortgage Default Data

*** Generalized Linear Model ***

Call: glm(formula = EVER90 ~ FICO + LTV + INCOME + CA, family = binomial(link = logit), data = credit, na.action = na.exclude, control = list(epsilon = 0.0001, maxit = 50, trace = F))

Deviance Residuals:

Min	1Q	Median	3Q	Max
–1.523824	–0.536522	–0.2340431	0.3169709	2.565638

Coefficients:

	Value	Std. Error	t value
(Intercept)	19.07413721	5.564345576	3.427921
FICO	–0.03189893	0.007279867	–4.381802
LTV	5.69435393	2.800980222	2.032986
INCOME	–0.40185724	0.246608000	–1.629539
CA	0.57827914	0.389452398	1.484852

(Dispersion Parameter for Binomial family taken to be 1)

Null Deviance: 118.4944 on 94 degrees of freedom

Residual Deviance: 64.35339 on 90 degrees of freedom

Number of Fisher Scoring Iterations: 5

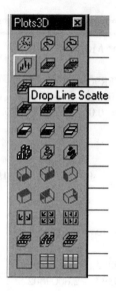

Figure 5.4   3D Graphic palette showing Drop Line scatter choice.

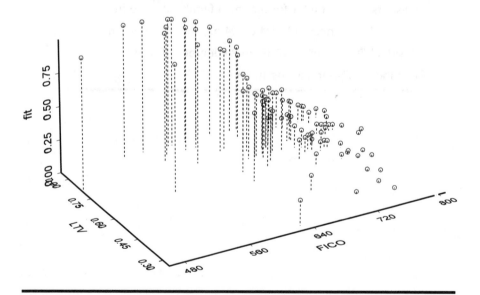

Figure 5.5   Plot of probability of mortgage default against FICO and LTV.

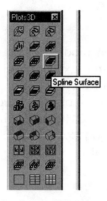

Figure 5.6   3D Graphic palette showing Spline Surface choice.

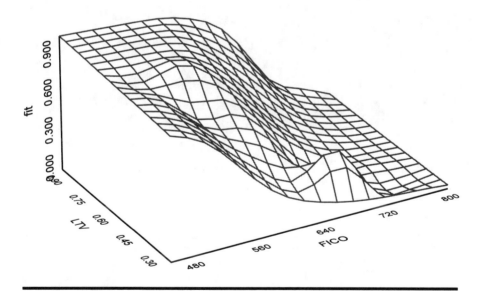

Figure 5.7   Spline surface for probability of mortgage default as a function of the variables FICO and LTV.

Exercises

5.1. In a logistic regression model for the GHQ data, which includes both the GHQ score and sex as explanatory variables, what would be the interpretation of the regression coefficient for sex?

5.2. Find the standard errors of the fitted values from the logistic regression model for the GHQ data, which includes only the single explanatory variable **ghqscore**. Plot the fitted values showing the 95% confidence interval.

Chapter 6

Analysing Longitudinal Data: Beating the Blues

6.1 Description of Data

The data to be used in this chapter arise from a study of an interactive, multimedia program designed to deliver *cognitive behavioural therapy* (CBT) to depressed patients via a computer terminal; the data are shown in Table 6.1. Patients were randomly allocated to receive either the computer therapy program, which was known as Beat the Blues (BtB), or treatment as usual (TAU). The trial protocol specified recording a measure of depression prior to the beginning of treatment and then again 2, 3, 5, and 8 months after the start of the treatment. Note that some patients in the trial 'drop-out' and do not have a full set of four post-randomisation depression recordings. (I am grateful to Dr. Judy Proudfoot for allowing me to use these data.)

The main question of interest about these data is whether or not the new computer approach for delivery of treatment for depression is effective or not.

6.2 Analysing Longitudinal Data

Longitudinal data arise when participants in a study are measured on the same variable on several different occasions. The data in Table 6.1 are an example in which depression is measured on five occasions, one before

Table 6.1   Beat the Blues Data

group	pre	m2	m3	m5	m8
0	25	12	5	7	7
0	13	12	11	11	10
0	17	17	15	14	14
1	20	20	16	15	12
1	23	12	4	6	5
1	42	10	8	10	6
1	16	10	7	9	2
0	18	12	11	10	NA
1	15	7	4	3	5
1	14	14	6	5	5
0	19	16	6	5	3
0	35	15	10	10	11
1	41	8	5	9	4
1	28	18	17	NA	NA
1	24	5	5	6	6
1	19	4	7	1	2
1	26	21	12	8	4
0	19	18	10	10	8
0	15	15	18	15	15
0	31	18	8	8	7
1	19	4	3	3	3
1	36	14	7	1	1
0	13	15	NA	NA	NA
1	35	12	10	8	10
0	26	29	25	23	20
0	19	18	23	23	23
1	16	1	0	0	NA
1	13	5	3	3	0
1	18	10	10	6	8
0	17	19	14	NA	NA
0	15	25	20	20	20
0	21	22	14	13	12
1	19	3	3	3	1
1	18	4	4	9	NA
0	22	10	13	12	14
0	15	12	5	5	4
0	21	15	17	8	13
0	16	15	14	15	12
0	22	22	22	23	24
1	13	10	6	2	4

treatment begins and four after the start of treatment. Analysis of such data has become something of a growth industry in statistics, largely because of its increasing importance in clinical trials (see Everitt and Pickles, 2000). There are a number of possible approaches to the analysis of such data:

- *Time-by-time analysis*: A series of *t*-tests are used to test for differences between the two groups at each time point. (In examples with more than two groups, a series of one-way analyses of variance might be used.) The procedure is straightforward but has a number of serious flaws and weaknesses that are detailed in Everitt (2001); consequently, it will not be pursued further here.
- *Response feature analysis — the use of summary measures*: Here the repeated measures for each participant are transformed into a single number considered to capture some important aspect of the participant's response profile. The summary measure to use has to be decided on before analysis of the data begins and must, of course, be relevant to the particular questions that are of interest in the study. Commonly used summary measures include
  1. Overall mean
  2. Maximum (minimum) value
  3. Time to maximum (minimum) response
  4. Slope of regression line on time
  5. Time to reach a particular value (for example, a fixed percentage of baseline)

  Having identified a suitable summary measure, a simple *t*-test (or analysis of variance) can be applied to assess between group differences.
- *Random effects model*: A detailed analysis of longitudinal data requires consideration of models that represent both the level and the shape of a group's profile of repeated measurements and also accounts adequately for the observed pattern of dependencies in these measurements. A flexible way to do this is to use regression-type models that include random effects for subjects, the presence of which allows for particular patterns of covariance between the repeated measures. For example, for the data in Table 6.1 we might first postulate a simple *random intercept model* of the form

$$y_{ijk} = \left(\beta_0 + \alpha_k\right) + \beta_1 \text{Time}_j + \beta_2 \text{Group}_i + \epsilon_{ijk} \tag{6.1}$$

where y_{ijk} represents the depression score for the kth patient on the jth post-randomisation occasion (Time$_j$, taking values 2, 3, 5, or 8) in the ith treatment group (Group$_i$, a dummy variable coding group membership),

β_0, β_1, and β_2 are regression coefficients, and the ϵ_{ijk} are error terms assumed to be normally distributed with zero mean and variance σ^2. The α_k are the patient random effect terms that model the shift in intercept for each subject; these are also assumed to be normally distributed, again with zero mean but variance σ_a^2. Such a model implies *compound symmetry* for the covariance matrix of the repeated measures, i.e., equality of the correlation between each pair of repeated measures (see Everitt and Dunn, 2001, for details). Other covariates such as pre-treatment depression score could be included in the model if necessary. Random effects models are described in detail in Pinheiro and Bates (2000).

6.3 Analysis Using S-PLUS

We shall begin by using the summary measure approach in the depression data in Table 6.1, and then move on to consider the fitting of random effects models to the data. For both, we shall concentrate on the command language approach.

6.3.1 Summary Measure Analysis of the Depression Data

The data are available in the **depress** data frame as shown in Table 6.2. As the summary measure for these data we shall simply use the mean of the post-randomisation measures. To begin, we shall simply ignore the pretreatment depression score available for each patient.

One question that needs to be addressed before calculating the chosen summary measure for each patient is what to do about those patients who drop out before the 8-month time point and, consequently, do not have all four intended post-randomisation depression measures. There are three possibilities:

1. Consider only patients who *do* have all post-randomisation measures,
2. Use all patients and calculate the mean of their *available* post-randomisation measures.
3. *Impute* (estimate) the missing values in some way; for example, use the *last observation carried forward* approach (LOCF), in which the missing values are replaced by the last value recorded for a patient.

For reasons explained in Everitt and Pickles (2000), the second option is the one generally recommended, and it is the one we will use here (but see Exercise 6.1).

Table 6.2   Contents of the Depress Data Frame

> depress

	subject	group	pre	m2	m3	m5	m8
1	1	TAU	25	12	5	7	7
2	2	TAU	13	12	11	11	10
3	3	TAU	17	17	15	14	14
4	4	BtB	20	20	16	15	12
5	5	BtB	23	12	4	6	5
6	6	BtB	42	10	8	10	6
7	7	BtB	16	10	7	9	2
8	8	TAU	18	12	11	10	NA
9	9	BtB	15	7	4	3	5
10	10	BtB	14	14	6	5	5
11	11	TAU	19	16	6	5	3
12	12	TAU	35	15	10	10	11
13	13	BtB	41	8	5	9	4
14	14	BtB	28	18	17	NA	NA
15	15	BtB	24	5	5	6	6
16	16	BtB	19	4	7	1	2
17	17	BtB	26	21	12	8	4
18	18	TAU	19	18	10	10	8
19	19	TAU	15	15	18	15	15
20	20	TAU	31	18	8	8	7
21	21	BtB	19	4	3	3	3
22	22	BtB	36	14	7	1	1
23	23	TAU	13	15	NA	NA	NA
24	24	BtB	35	12	10	8	10
25	25	TAU	26	29	25	23	20
26	26	TAU	19	18	23	23	23
27	27	BtB	16	1	0	0	NA
28	28	BtB	13	5	3	3	0
29	29	BtB	18	10	10	6	8
30	30	TAU	17	19	14	NA	NA
31	31	TAU	15	25	20	20	20
32	32	TAU	21	22	14	13	12
33	33	BtB	19	3	3	3	1
34	34	BtB	18	4	4	9	NA
35	35	TAU	22	10	13	12	14
36	36	TAU	15	12	5	5	4
37	37	TAU	21	15	17	8	13
38	38	TAU	16	15	14	15	12
39	39	TAU	22	22	22	23	24
40	40	BtB	13	10	6	2	4

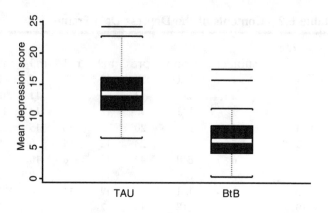

Figure 6.1 Box plots of mean summary measure for Beat the Blues data.

To begin, we shall calculate the chosen summary measure for each patient in the study and then use a box plot to display the measures graphically for each treatment group.

```
attach(depress)
depress.sm<-apply(depress[,4:7],1,mean,na.rm=T)
#use the apply function to get the mean of available
#post-randomization measurements for each patient
boxplot(depress.sm[group=="TAU"],depress.sm[group=="BtB"],
names=c("TAU","BtB"),ylab="Mean depression score")
```

The resulting diagram is shown in Figure 6.1. This suggests that the depression score is higher for the subjects receiving treatment as usual; this can be tested formally with an independent samples *t*-test.

```
t.test(depress.sm[group=="TAU"],depress.sm[group=="BtB"])
```

The results are shown in Table 6.3. There is a very significant difference between treatment groups, with the average depression score for TAU being between 4 and 10 points higher than that of the BtB group.

We can take account of the pretreatment depression value in the summary measure approach by using the **lm** function to model the summary measure in terms of the pretreatment value and the treatment group.

```
summary(lm(depress.sm~pre+group))
```

The corresponding results are shown in Table 6.4. Since the default coding used by S-PLUS for the two levels of group is −1 for TAU and +1

Table 6.3   Results of Applying a *t*-Test to the Mean Summary Measure of Post-Randomisation Depression Values

> t.test(depress.sm[group == "TAU"], depress.sm[group == "BtB"])

    Standard Two-Sample t-Test

data: depress.sm[group == "TAU"] and depress.sm[group == "BtB"]
t = 4.91, df = 38, p-value = 0
alternative hypothesis: true difference in means is not equal to 0
95 percent confidence interval:
      4.27   10.26
sample estimates:
  mean of x mean of y
      14.2   6.94

Table 6.4   Analysis of Mean Depression Score Taking Account of Pre-Randomisation Depression Value

> summary(lm(depress.sm ~ pre + group))

Call: lm(formula = depress.sm ~ pre + group)
Residuals:

Min	1Q	Median	3Q	Max
−7.231	−3.032	−0.667	1.929	10.06

Coefficients:

	Value	Std. Error	t value	Pr(>|t|)
(Intercept)	8.5452	2.2530	3.7928	0.0005
pre	0.0948	0.0997	0.9511	0.3477
group	−3.7639	0.7542	−4.9907	0.0000

Residual standard error: 4.688 on 37 degrees of freedom
Multiple R-Squared: 0.4023
F-statistic: 12.45 on 2 and 37 degrees of freedom, the p-value
  is 0.00007319

Correlation of Coefficients:

	(Intercept)	pre
pre	−0.9443	
group	0.1747	−0.1850

for BtB, the regression coefficient for group and its standard error need to be doubled to give the estimated treatment effect adjusted for pretreatment depression, and the appropriate confidence interval for the effect. Consequently, a 95% confidence interval for treatment effect is (−7.53 − 1.96 × 1.51, −7.53 + 1.96 × 1.51), i.e., (−10.50, −4.57). Beat the Blues is estimated to reduce the depression score by between 10 and 5 points on average. Pretreatment score appears to have little effect on the scores post-treatment.

6.3.2 Random Effects Models for the Depression Data

Before undertaking the formal modelling of the depression data, it may be useful to consider some informative graphical displays. According to Diggle et al. (1994), there is no single prescription for making effective graphical displays of longitudinal data, although they do offer the following simple guidelines:

- Show as much of the relevant raw data as possible, rather than only data summaries.
- Highlight aggregate patterns of potential scientific interest.
- Identify both cross-sectional and longitudinal patterns.
- Make easy identification of unusual individuals or unusual observations.

Two graphical displays of the depression data that meet some of these requirements will now be constructed using the S-PLUS command language, the first a plot of individual patient profiles and the second a plot of group means and standard deviations:

```
lt<-rep(1,40)
lt[group=="BtB"]<-2
#set up vector lt with values 1 for TAU and 2 for BtB
matplot(c(0,2,3,5,8),t(depress[,3:7]),type="l",
lty=lt,axes=F,xlab="Visit",ylab="Depression",
col=1)
#use the matplot function to plot each patient's profile
#of depression values, setting the line type to 1 for TAU
#and to 2 for BtB using the previously constructed vector
#lt. Labels but does not draw x or y axes
axis(1,at=c(0,2,3,5,8),labels=c("pre","m2",
"m3","m5","m8"))
```

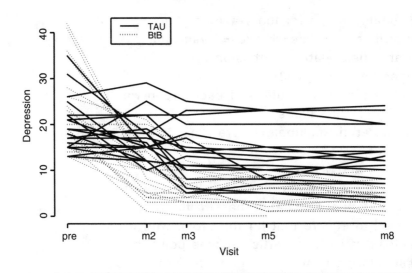

Figure 6.2 Individual participant profiles for the Beat the Blues data identifying treatment group.

```
#adds x-axis with tick marks at appropriate
#points and suitably labelled
axis(2)
legend(locator(1),c("TAU","BtB"),lty=1:2)
#adds y-axis and a legend interactively
```

The resulting plot is shown in Figure 6.2. The tendency for the profiles of the patients given the BtB treatment to be lower throughout the post-randomisation visits is clearly visible.

Now let us consider constructing a plot of the group means with standard deviation bars shown:

```
m1<-apply(depress[group=="TAU",3:7],2,mean,na.rm=T)
s1<-
sqrt(apply(depress[group=="TAU",3:7],2,var,na.action="omit")
m2<-apply(depress[group=="BtB",3:7],2,mean,na.rm=T)
s2<-sqrt(apply(depress[group=="BtB",3:7],2,var,
na.action="omit"))
#calculate means and standard deviations of
#each group
times <-c(0,2,3,5,8)
ylim<-range(m1+s1,m2+s2,m1-s1,m2-s2)
#store values for x-axis in times and a
```

```
#suitable range for the y-axis in ylim
plot(times,m1,type="l",axes=F,ylim=ylim,
xlab="Visit",ylab="Depression")
lines(times,m2,lty=2)
#plots the mean profiles of each group with
#different line types
axis(1,at=times,labels=c("pre","m2","m3","m5","m8"))
axis(2)
#add x and y axes
#now the segments function will be used to
#add standard deviation bars to the plot
#need to ensure that for the group with the highest
#mean value at a particular time point to
#standard deviation line goes upwards, and
#for the group with the lower mean downwards.
pm<-compare(m1,m2)
m1s1<-m1+pm*s1
m2s2<-m2-pm*s2
#compare returns a vector of +1, -1, and 0s
#depending on the differences between the corresponding
#elements of m1 and m2
segments(times,m1,times,m1s1)
segments(times−0.1,m1s1,times+0.1,m1s1)
segments(times,m2,times,m2s2)
segments(times−0.1,m2s2,times+0.1,m2s2)
legend(locator(1),c("TAU","BtB"),lty=1:2)
```

The resulting diagram is shown in Figure 6.3. On average, the depression values for patients given the BtB treatment are lower on each post-randomisation visit than those given treatment as usual.

Now we move on to consider the fitting of random effect models to the depression data using the **lme** function. The first step is to rearrange the data so that the values of the four post-randomisation depression measures of each patient are stacked into one long vector, and then the values of the subject identifier, time of measurement, treatment group, and pretreatment depression need to be replicated in accordance with this new arrangement of the response variable. Suitable commands are as follows:

```
Depression<-as.vector(as.matrix(t(depress[,4:7)))
#stacks the repeated depression values into a single vector
```

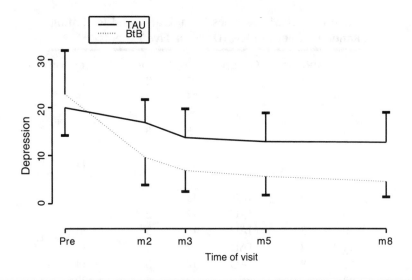

Figure 6.3 Plot of mean profiles and standard deviations for each treatment group in the Beat the Blues data.

```
#after transposing using the t function
Group<-rep(group,rep(4,40))
Pre<-rep(pre,rep(4,40))
Subject<-rep(subject,rep(4,40))
Time<-rep(c(2,3,5,8),40)
depress.new<-data.frame(Subject,Group,Pre,Time, Depression)
#the rep function is used to replicate the
#explanatory variables, subject number and
#time of measurement appropriately and then the
#rearranged values are combined into a new
#data frame.
```

The first 20 rows of **depress.new** are shown in Table 6.5.

To begin, we shall fit the random effects model described by Equation 6.1 with the addition of the pretreatment depression value as an extra covariate.

```
fit1<-lme(Depression~Pre+Group+Time,method="ML",
random=~1|Subject,data=depress.new,na.action=
na.omit)
#Depression is modelled in terms of the pretreatment
#value, treatment group, and time of measurement
```

**Table 6.5 Beat the Blues Data Rearranged for Fitting
Random Effect Models (Data for Five Patients)**

	Subject	Group	Pre	Time	Depression
1	1	TAU	25	2	12
2	1	TAU	25	3	5
3	1	TAU	25	5	7
4	1	TAU	25	8	7
5	2	TAU	13	2	12
6	2	TAU	13	3	11
7	2	TAU	13	5	11
8	2	TAU	13	8	10
9	3	TAU	17	2	17
10	3	TAU	17	3	15
11	3	TAU	17	5	14
12	3	TAU	17	8	14
13	4	BtB	20	2	20
14	4	BtB	20	3	16
15	4	BtB	20	5	15
16	4	BtB	20	8	12
17	5	BtB	23	2	12
18	5	BtB	23	3	4
19	5	BtB	23	5	6
20	5	BtB	23	8	5

```
#the random argument specifies a single random
#effect for each 'group' of response values
#the 'group' here being subject.
summary(fit1)
```

The results are shown in Table 6.6. Both the treatment group and time-fixed effects are very significant whilst the pretreatment score effect is again nonsignificant. The estimated standard deviation of the random intercept effects is 4.226 and the corresponding value for the residual terms is 2.738.

Now consider a model with a random intercept and random slope for each patient:

```
fit2<-lme(Depression~Pre+Group+Time,method="ML",
random~Time|Subject,data=depress.new,na.action =
na.omit)
summary(fit2)
```

Table 6.6 Results of Fitting Random Intercept Model to Beat the Blues Data

```
> summary(fit1)
Linear mixed-effects model fit by maximum likelihood
 Data: depress.new
    AIC        BIC       logLik
  830.8885   848.9523   –409.4443
```

Random effects:
 Formula: ~ 1 | Subject

	(Intercept)	Residual
StdDev:	4.22594	2.737807

Fixed effects: Depression ~ Pre + Group + Time

	Value	Std.Error	DF	t-value	p-value
(Intercept)	11.23785	2.215434	109	5.072525	<.0001
Pre	0.10093	0.096079	37	1.050468	0.3003
Group	–3.76292	0.727953	37	–5.169180	<.0001
Time	–0.66767	0.102413	109	–6.519347	<.0001

Correlation:

	(Intr)	Pre	Group
Pre	–0.925		
Group	0.164	–0.178	
Time	–0.186	–0.012	–0.001

Standardized Within-Group Residuals:

Min	Q1	Med	Q3	Max
–1.825689	–0.5614503	–0.07627247	0.5261911	3.095711

Number of Observations: 150
Number of Groups: 40

The results are given in Table 6.7. The tests for the fixed effects in the model have results very similar to those for the random intercept model as do the estimated standard deviations for the random intercept terms and the residual terms. The estimated standard deviation of the random slope effects is 0.363.

Table 6.7   Results of Fitting Random Intercept and Random Slope Model to Beat the Blues Data

```
> summary(fit2)
Linear mixed-effects model fit by maximum likelihood
 Data: depress.new
     AIC        BIC        logLik
  833.3486   857.4337   –408.6743

Random effects:
 Formula: ~ Time | Subject
 Structure: General positive-definite
                StdDev       Corr
(Intercept)   4.4034190    (Inter
       Time   0.3632755    –0.277
   Residual   2.5745429
```

Fixed effects: Depression ~ Pre + Group + Time

	Value	Std.Error	DF	t-value	p-value
(Intercept)	11.16996	2.215142	109	5.042547	<.0001
Pre	0.10368	0.095911	37	1.081000	0.2867
Group	–3.73918	0.726462	37	–5.147112	<.0001
Time	–0.66474	0.114392	109	–5.811035	<.0001

Correlation:

	(Intr)	Pre	Group
Pre	–0.924		
Group	0.164	–0.179	
Time	–0.196	–0.011	–0.002

Standardized Within-Group Residuals:

Min	Q1	Med	Q3	Max
–1.938337	–0.5395204	–0.1078871	0.4987325	2.795758

```
Number of Observations: 150
Number of Groups: 40
```

We can assess whether the random intercept and random slopes model fits better than the simpler random intercepts model using the **anova** function.

```
anova(fit1,fit2)
```

Table 6.8   Comparing Random Effects Models

	Model	df	AIC	BIC	logLik	Test	L.Ratio	p-value
> anova(fit1, fit2)								
fit1	1	6	830.8885	848.9523	−409.4443			
fit2	2	8	833.3486	857.4337	−408.6743	1 vs 2	1.53987	0.463

Table 6.9   Confidence Intervals for Parameters in Random Intercept Model for Beat the Blues Data

> intervals(fit1)
Approximate 95% confidence intervals

Fixed effects:

	lower	est.	upper
(Intercept)	6.90586774	11.2378455	15.5698233
Pre	−0.09113376	0.1009282	0.2929902
Group	−5.21809383	−3.7629201	−2.3077463
Time	−0.86791956	−0.6676652	−0.4674108

Random Effects:
Level: Subject

	lower	est.	upper
sd((Intercept))	2.987841	4.22594	5.977081

Within-group standard error:

lower	est.	upper
2.27141	2.737807	3.29997

giving the results in Table 6.8. Clearly, the more-complicated model does not represent an improvement in this case.

We can find confidence intervals for the parameters of the random intercept model using the **intervals** function.

    intervals(fit1)

gives the confidence intervals shown in Table 6.9. (Remember that the confidence interval given for the treatment group is based on the −1,1 coding used by default.)

Now we need to examine the residuals from the fitted model to check the various assumptions made. A plot of standardised residuals against

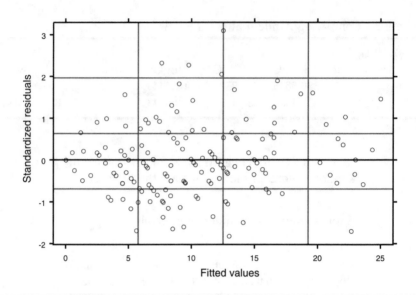

Figure 6.4 Plot of standardised residuals from random intercept model for Beat the Blues data.

fitted values, for example, can be used to assess the assumption of constant variance of the ϵ_{ij}:

```
plot(fit1)
```

The plot is shown in Figure 6.4. In this plot we are looking for a systematic increase (or, less commonly, a systematic decrease) in the variance of the residuals as the fitted values increase. If this is present, the residuals on the right-hand side of the plot will have a greater vertical spread than those on the left, forming a horizontal wedge-shaped pattern. There is no such pattern in Figure 6.4.

The assumption of normality for the error terms can be assessed by a normal probability plot of the residuals produced by the **qqnorm** function:

```
qqnorm(fit1,~resid(.)|Group)
#the second argument specifies the residuals
#from the fitted model and plotting by
#treatment group
```

The resulting plot is shown in Figure 6.5. There is little evidence of a departure from normality in either group.

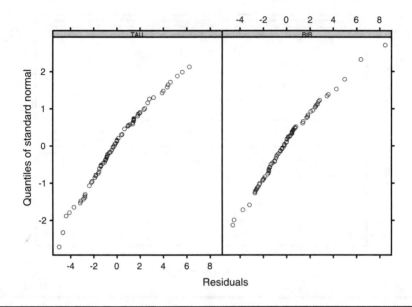

Figure 6.5 Normal probability plot of standardized residuals from a random intercept model fitted to the Beat the Blues data.

Lastly, we can assess the normality assumption of the estimated random effects (how these are estimated is described in Pinheiro and Bates, 2000), again by using the **qqnorm** function:

```
qqnorm(fit1,~ranef(.)|Group)
```

giving Figure 6.6. Again, there is no evidence of any serious departure from the assumption.

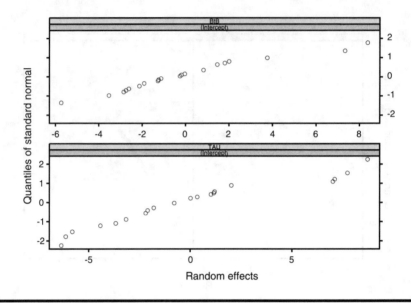

Figure 6.6 Normal probability plot of estimated random effects in a random intercepts model fitted to the Beat the Blues data.

Exercises

6.1. The box plot shown in Figure 6.1 suggests that the mean summary measure does not have the same variance in each treatment group and, consequently, violates at least one of the assumptions made by the *t*-test applied in the text. Investigate the use of the *t*-test function in a version that does not make this homogeneity of variance assumption.

6.2. Apply the mean summary measure approach to the depression data after excluding all patients who do not have all four intended post-randomisation depression values.

6.3. Apply the mean summary measure approach to the depression data after replacing missing values with the last observation carried forward.

Chapter 7

Nonlinear Regression and Maximum Likelihood Estimation: Athletes and Geysers

7.1 Description of Data

In this chapter we shall examine the functions available in S-PLUS for numerical optimisation in the context of both *nonlinear regression* and *maximum likelihood estimation*. To illustrate the former we shall use the data shown in Table 7.1, which show times in seconds recorded by the winners of the men's Olympic 1500-m event from 1900 to 2000 (there were no Olympic Games in 1916, 1940, and 1944). Interest here will centre on fitting a particular regression model to the data that might be used to predict future winning times, and the ultimate time achievable.

As an example of the application of maximum likelihood estimation in S-PLUS we shall estimate the parameters of a *normal mixture distribution* (see Everitt and Hand, 1981) fitted to the data in Table 7.2. These data consist of the waiting times between successive eruptions of the Old Faithful geyser in Yellowstone National Park, Wyoming, USA. There were 300 eruptions observed in the period of observation (August 1 to 15, 1985), so Table 7.2 contains 299 waiting times measured in minutes.

Table 7.1 Olympic 1500-m Winning Times (men)

Year	Time
1900	246.00
1904	245.40
1908	243.40
1912	236.80
1920	241.80
1924	233.60
1928	233.20
1932	231.20
1936	227.80
1948	225.20
1952	225.20
1956	221.20
1960	215.60
1964	218.10
1968	214.90
1972	216.30
1976	219.20
1980	218.40
1984	212.50
1988	215.96
1992	220.12
1996	215.78
2000	212.07

7.2 Nonlinear Regression and Maximum Likelihood Estimation

An examination of the 1500-m times shows that they are improving (decreasing) over time. From physiological consideration it is clear that there is a lower limit to the times. Such considerations led Chatterjee and Chatterjee (1982) to suggest the *three-parameter exponential* for modelling this type of athletic records data. The model is as follows:

$$E(\text{time}) = \theta_1 + \theta_2 \exp\left[\theta_3\left(\text{Year} - 1900\right)\right] \qquad (7.1)$$

All three parameters in the model have meaningful interpretations:

Table 7.2   Waiting Times (minutes) for Eruption of Old Faithful

80	71	57	80	75	77	60	86	77	56	81	50	89	54	90	73
60	83	65	82	84	54	85	58	79	57	88	68	76	78	74	85
75	65	76	58	91	50	87	48	93	54	86	53	78	52	83	60
87	49	80	60	92	43	89	60	84	69	74	71	108	50	77	57
80	61	82	48	81	73	62	79	54	80	73	81	62	81	71	79
81	74	59	81	66	87	53	80	50	87	51	82	58	81	49	92
50	88	62	93	56	89	51	79	58	82	52	88	52	78	69	75
77	53	80	55	87	53	85	61	93	54	76	80	81	59	86	78
71	77	76	94	75	50	83	82	72	77	75	65	79	72	78	77
79	75	78	64	80	49	88	54	85	51	96	50	80	78	81	72
75	78	87	69	55	83	49	82	57	84	57	84	73	78	57	79
57	90	62	87	78	52	98	48	78	79	65	84	50	83	60	80
50	88	50	84	74	76	65	89	49	88	51	78	85	65	75	77
69	92	68	87	61	81	55	93	53	84	70	73	93	50	87	77
74	72	82	74	80	49	91	53	86	49	79	89	87	76	59	80
89	45	93	72	71	54	79	74	65	78	57	87	72	84	47	84
57	87	68	86	75	73	53	82	93	77	54	96	48	89	63	84
76	62	83	50	85	78	78	81	78	76	74	81	66	84	48	93
47	87	51	78	54	87	52	85	58	88	79					

- θ_1 is the ultimate limit.
- θ_2 is the amount by which the first record (i.e., for year 1900) exceeds the ultimate limit.
- θ_3 is the constant proportional rate of decline in the record series towards the ultimate limit.

The three parameters can be estimated by minimizing the sum of squares of the differences between the observed values of the 1500-m times and those predicted by the assumed model.

For the geyser eruption data we shall investigate whether there is any evidence of multimodality in the data by fitting a mixture of two normal densities, i.e., the density function given by

$$f(x) = N(x;\mu_1,\sigma_1) + (1-p)N(x;\mu_2,\sigma_2) \tag{7.2}$$

where

$$N(x;\mu_i,\sigma_i) = \frac{1}{\sigma_i\sqrt{2\pi}} e^{-\frac{1}{2}\left(\frac{x-\mu_i}{\sigma_i}\right)^2}$$

In this density there are five parameters to estimate: the mixing proportion, p, and the means, μ_1 and μ_2, and standard deviations, σ_1 and σ_2, of each component normal density. If such a density function does provide a good description of the eruption data, it implies something fundamental about the mechanism of the eruption process.

To estimate the five parameters in Equation 7.2 we shall use maximum likelihood, maximizing the log-likelihood of n observations $x_1, \ldots, x_n$ from $f(x)$, in (7.2) i.e.,

$$ l = \sum_{i=1}^{n} \log\left\{ N\left(x_i; \mu_1, \sigma_1\right) + \left(1 - p\right) N\left(x_i; \mu_2, \sigma_2\right)\right\} \qquad (7.3) $$

7.3 Analysis Using S-PLUS

Both the nonlinear regression and maximum likelihood estimation described in the previous section require the numerical optimisation of some specified function. In S-PLUS there are a number of functions available for such optimisation; here we illustrate the use of two of these, **nls** and **nlminb**.

7.3.1 Modelling the Olympic 1500-m Times

To estimate the three parameters in the nonlinear model specified in Equation 7.1 we shall use the **nls** function in S-PLUS. This function estimates parameters by minimizing the sum of squares of differences between the response and the model prediction. The data are available in an S-PLUS data frame **Olympic**. To begin, we shall store initial values of the three parameters, θ_1, θ_2, and θ_3 along with this data frame.

```
param(Olympic,"theta1") <-200
param(Olympic,"theta2") <-40
param(Olympic,"theta3")<- -0.01
#these starting values are suggested by the
#results in Chatterjee and Chatterjee, 1982.
```

The contents of the amended version of **Olympic** are shown in Table 7.3. We can now estimate the parameters using the **nls function**:

```
Olympic.fit<-nls(Time~theta1+theta2*exp(theta3*(Year —
1900)),Olympic)
#Formula used to specify the nonlinear model
Olympic.fit$parameters
#this object contains final values of parameters.
```

**Table 7.3 Contents of Olympic
Data Frame**

Parameters:
$theta1:
[1] 200

$theta2:
[1] 40

$theta3:
[1] -0.01

Variables:

	Year	Time
1	1900	246.00
2	1904	245.40
3	1908	243.40
4	1912	236.80
5	1920	241.80
6	1924	233.60
7	1928	233.20
8	1932	231.20
9	1936	227.80
10	1948	225.20
11	1952	225.20
12	1956	221.20
13	1960	215.60
14	1964	218.10
15	1968	214.90
16	1972	216.30
17	1976	219.20
18	1980	218.40
19	1984	212.50
20	1988	215.96
21	1992	220.12
22	1996	215.78
23	2000	212.07

The results are shown in Table 7.4. To obtain parameter estimates using a different set of initial values we need to use the **start** argument of **nls**, for example,

Table 7.4   Final Parameter Estimates for Model Fitted to 1500-m Times Using Two Sets of Starting Values

(1) **Starting values 1: θ_1 = 200, θ_2 = 40, θ_3 = –0.01**

	theta1	theta2	theta3
Estimated values	206.0838	41.86718	–0.01733334

(2) **Starting values 2: θ_1 = 180, θ_2 = 30, θ_3 = –0.02**

	theta1	theta2	theta3
Estimated values	206.0941	41.85928	–0.01734269

```
Olympic.fit1<-nls(Time~theta1+theta2*exp(theta3*(Year —
1900)),start = list(theta1 = 180,theta2 = 30,theta3 = -0.02),
Olympic)
Olympic.fit1$parameters
```

The results obtained with these starting values are also shown in Table 7.4. They are almost identical to those obtained with the first set of starting values.

We can now examine the fit of the model graphically by plotting the original observations and adding the predicted values from the fitted model.

```
attach(Olympic)
plot(Year,Time)
lines(Year,predict(Olympic.fit))
```

The graph is shown in Figure 7.1. The fit appears to be very good.

One way to obtain confidence intervals for the estimated parameters is to use the *bootstrap* approach, which is described in detail in Efron and Tibshirani (1993), but which consists essentially of repeated sampling of the data with replacement, followed by calculation of the parameters for each such *bootstrap sample*. In this way empirical distributions of the parameters can be constructed from which confidence intervals can be determined from the appropriate quantiles, i.e., 2.5 and 97.5% for a 95% confidence interval. In S-PLUS bootstrapping is available by using the **bootstrap** function

```
set.seed(962)
#to ensure readers have same set of random samples
Olympic.boot<-bootstrap(Olympic,nls(Time~theta1+
theta2*exp(theta3*(Year — 1900)),Olympic)$parameters, B =
1000)
#we use 1000 bootstrap samples; starting
```

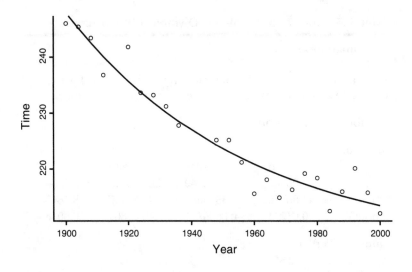

Figure 7.1 **Plot of Olympic 1500-m times and three-parameter exponential prediction.**

```
#values in each case will be the parameter
#values stored with the Olympic data frame
summary(Olympic.boot)
plot(Olympic.boot)
```

The numerical results are shown in Table 7.5 and the constructed distributions of the parameters are shown graphically in Figure 7.2. The empirical percentiles lead to 95% confidence intervals as follows:

θ_1: [192.79, 213.21]
θ_2: [35.59, 56.67]
θ_3: [−0.027, −0.011]

The BCa (*bias corrected accelerated*) confidence intervals, which are based on an improved procedure described in Efron and Tibshirani (1993), are very similar.

7.3.2 Estimating the Parameters in a Mixture Fitted to the Geyser Eruption Data

Here we shall use the **nlminb** function to estimate the parameters in a mixture of two normal densities fitted to the geyser eruption data, using maximum likelihood. First we need to write an S-PLUS function to evaluate

Table 7.5   Bootstrap Results for Olympic 1500-m Times

> summary(boot)
Call:
bootstrap(data = Olympic, statistic = nls(Time ~ theta1 + theta2 *
exp(theta3 * (Year - 1900)), Olympic)$parameters, B = 1000)

Number of Replications: 1000

Summary Statistics:

	Observed	Bias	Mean	SE
theta1	206.08380	−0.6986649	205.38513	5.355482
theta2	41.86718	1.3503861	43.21757	5.764726
theta3	−0.01733	−0.0006773	−0.01801	0.004886

Empirical Percentiles:

	2.5%	5%	95%
theta1	192.79244	196.08406	212.39022
theta2	35.59040	36.70991	52.87415
theta3	−0.02711	−0.02495	−0.01199

BCa Confidence Limits:

	2.5%	5%	95%	97.5%
theta1	194.23732	196.88147	212.69860	213.82443
theta2	34.71046	35.72474	50.23496	53.32543
theta3	−0.02608	−0.02399	−0.01171	−0.01062

Correlation of Replicates:

	theta1	theta2	theta3
theta1	1.0000	−0.7339	−0.7976
theta2	−0.7339	1.0000	0.2300
theta3	−0.7976	0.2300	1.0000

the log-likelihood for any values of the five parameters. We shall call this function **LL**.

```
LL<-function(params,data){
#params contains the five parameter values of
#the mixture density in the order, p, mu1,
#sigma1, mu2, sigma2, and data contains
#the observed values — in our example the
#eruption waiting times
```

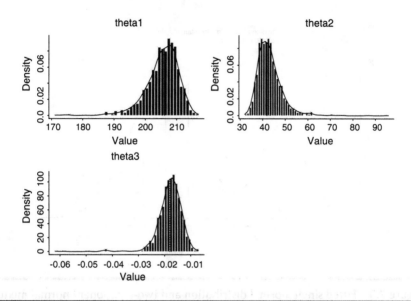

Figure 7.2 Bootstrap distributions of the three parameters in the three-parameter exponential distribution fitted to the Olympic 1500-m times (1000 bootstrap samples used).

```
t1<-dnorm(data,params[2],params[3])
t2<-dnorm(data,params[4],params[5])
#evaluate the two normal densities using the
#dnorm function
f<-params[1]*t1+(1-params[1])*t2
#evaluate the mixture density for all
#sample values
ll<-sum(log(f))
#evaluate the log-likelihood
-ll
#nlminb needs a function to minimize so
#we set the value returned by LL to
#minus the value of the log-likelihood
}
```

We assume the eruption waiting times are available in a vector **geyser**, and first we will need to examine a histogram of the data to get initial values for the five parameters in the mixture.

```
hist(geyser)
```

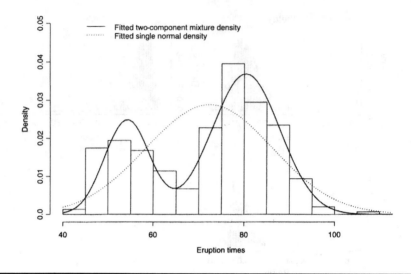

Figure 7.3   Fitted single normal distribution and two-component normal mixture for geyser eruption waiting times.

The resulting histogram is shown in Figure 7.3 (the fitted curves in the figure are explained later). Two modes are clearly visible, and reasonable starting values for the five parameters might be $p = 0.5$, $\mu_1 = 50$, $\sigma_1 = 10$, $\mu_2 = 80$, and $\sigma_2 = 10$.

Now we can fit the mixture distribution to the eruption waiting times, stored in the vector, **geyser**, using the **nlminb** function.

```
geyser.res<-nlminb(c(0.5,50,10,80,10),LL,
data = geyser,lower = c(0.0001,-Inf,0.0001,
-Inf,0.0001),upper = c(0.9999,Inf,Inf,Inf, Inf))
#starting values of p = 0.5, mu1 = 50, sigma1 = 10,
#mu2 = 80 and sigma2 = 10
#LL is the function to be minimized
#and data is the extra argument used
#by this function
#lower bound for p is set at 0.0001 and
#for sigma1 and sigma2 0.0001.
#an upper bound of 0.9999 is set for p
#all other bounds are set at infinity
```

This returns **geyser.res**, which is an S-PLUS list object containing a variety of information about the optimisation process. Of most interest are the values of the estimated parameters found from

```
geyser.res$parameters
```

giving

[1] 0.3075936 54.2026523 4.9520039 80.3603088 7.5076321

(Fewer decimal places could be obtained by using the **options** function, for example, **options(digits = 3)**.)

The fit of the mixture can be assessed by superimposing the fitted density onto a histogram of the observed eruption times.

```
x<-seq(40,120,length = 100)
#use the seq function to create a vector
#of 100 values between 40 and 120 at which
#to plot estimated mixture density
p<-geyser.res$parameters[1]
mu1<-geyser.res$parameters[2]
sig1<-geyser.res$parameters[3]
mu2<-geyser.res$parameters[4]
sig2<-geyser.res$parameters[5]
f<-pdnorm(x,mu1,sig1)+(1-p)dnorm(x,mu2,sig2)
#f contains fitted mixture densities corresponding
#to the values in x
hist(geyser,probability = T,col = 0,ylab = "Density",
ylim = c(0, 0.03),xlab = "Eruption waiting times")
#constructs histogram without colour or shading
#and as a density function
lines(x,f)
#adds estimated density to histogram
```

The resulting curves are shown in Figure 7.3. It is clear that the fitted mixture describes the eruption waiting times very well.

We can estimate the standard errors of the five parameters by again using a bootstrap approach. We first construct an S-PLUS function that returns the values of the five parameters for each bootstrap sample taken:

```
fit1<-function(data,start){
#data is a vector of sample values
#start is a vector of 5 starting values for
#the five parameters of the mixture; the mixing
#proportion is constrained to be in the interval 0.0001
```

```
#0.9999
data.res<-nlminb(start,LL,data = data,lower = c
0.0001,NA,0.0001,NA,0.0001),upper = c(
0.9999,NA,NA,NA,NA))
#estimate the five parameters
result <- data.res$parameters
result
}
```

Now the bootstrap can be applied; here we shall use 500 samples (be patient when running this portion of code — it takes some time!).

```
set.seed(9831)
geyser.boot<-bootstrap(geyser,fit1,args.stat =
list(start = c(0.05, 50, 10, 80, 10)),B = 500)
#takes 500 bootstrap samples and estimates
#the mixture parameters for each using the
#same starting values for each
summary(geyser.boot)
plot(boot)
```

The results are shown in Figure 7.4 and Table 7.6. In this case the estimated 95% confidence intervals for the five parameters are

> p: [0.245,0.365]
> μ_1: [52.500, 55.732]
> σ_1: [4.952, 0.527]
> μ_2: [78.994, 81.674]
> σ_2: [6.385, 8.792]

(Note that some samples cause problems for the estimation process because of the singularity problem described in Everitt and Hand, 1981; see also Exercise 7.3).

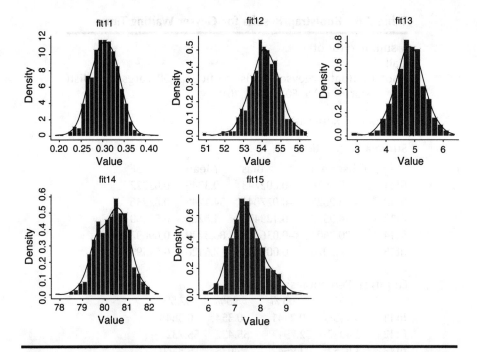

Figure 7.4   Bootstrap distributions for the five parameters in the two-component normal mixture fitted to the geyser eruption waiting time data; fit11 is the mixing proportion, fit12 is the mean of the first component, fit12 is the standard deviation of the first component, fit14 is the mean of the second component, and fit15 is the standard deviation of the second component.

Table 7.6   Bootstrap Results for Geyser Waiting Times

```
> summary(boot)
Call:
bootstrap(data = geyser, statistic = fit1, B = 500, args.stat = list(
      start = c(0.5, 50, 10, 80, 10)))
```

Number of Replications: 500

Summary Statistics:

	Observed	Bias	Mean	SE
fit11	0.3076	−0.002743	0.3049	0.03222
fit12	54.2027	−0.077846	54.1248	0.82236
fit13	4.9520	−0.129478	4.8225	0.52680
fit14	80.3603	−0.036408	80.3239	0.69097
fit15	7.5076	0.005208	7.5128	0.61591

Empirical Percentiles:

	2.5%	5%	95%	97.5%
fit11	0.2451	0.2541	0.354	0.3648
fit12	52.4997	52.7917	55.422	55.7321
fit13	3.8153	3.9642	5.675	5.8460
fit14	78.9938	79.2288	81.425	81.6744
fit15	6.3848	6.6201	8.654	8.7921

BCa Confidence Limits:

	2.5%	5%	95%	97.5%
fit11	0.2515	0.2606	0.3611	0.373
fit12	52.5957	52.8893	55.5269	55.765
fit13	4.0611	4.2227	5.9302	6.082
fit14	78.9484	79.2066	81.3629	81.539
fit15	6.6189	6.7132	8.7762	9.005

Correlation of Replicates:

	fit11	fit12	fit13	fit14	fit15
fit11	1.0000	0.4554	0.4573	0.3137	−0.3666
fit12	0.4554	1.0000	0.7401	0.4760	−0.4981
fit13	0.4573	0.7401	1.0000	0.4995	−0.5305
fit14	0.3137	0.4760	0.4995	1.0000	−0.4802
fit15	−0.3666	−0.4981	−0.5305	−0.4802	1.0000

Exercises

7.1. Investigate the residuals from fitting the model in Equation 7.1 to the Olympic records data. Is there any evidence of problems with the model?

7.2. Investigate the use of different numbers of bootstrap samples when finding confidence intervals for the parameters in the three parameter exponential model for the Olympic record data.

7.3. The mixture model in Equation 7.2 can give rise to singularity problems when fitted to data — essentially this means that the likelihood function tends towards infinity for reasons described in Everitt and Hand (1981). One solution is to constrain the two variances in the model to be the same. Fit such a model to the geyser eruption waiting times and construct a diagram showing a histogram of these times plus the two-fitted models (equal and unequal variances). Add an appropriate legend.

Chapter 8

Survival Analysis: Motion Sickness and Bird Survival

8.1 Description of Data

Two data sets will be used in this chapter. The first, shown in Table 8.1, arises from a research programme investigating motion sickness at sea. Human subjects were placed in a cubical cabin mounted on a hydraulic piston and subjected to vertical motion for 2 hours. The length of time until each subject first vomited was recorded. Some subjects requested an early stop to the experiment and some lasted the whole 2 hours without vomiting. (See Burns, 1984, for full details of the study.)

Our second data set is shown in Table 8.2. These data arise from 50 female black ducks from two locations in New Jersey that were captured and fitted with radios. The ducks were captured over a period of about 4 weeks and included 31 hatch-year birds (birds born during the previous breeding season) and 19 after-hatch birds (birds at least 1 year of age). The status (alive, missing, or dead) of each bird was recorded daily from the date of release until the end of the study. For some of the birds, death was not observed during the period of the study. Full details of the study are given in Pollock et al. (1989).

Both these data sets involve time to the occurrence of a particular event (vomiting in Table 8.1, death in Table 8.2). Such data are generally referred to by the generic term *survival data* even when the end point or event being considered is not death. Such data generally require special techniques for analysis for two main reasons:

Table 8.1 Data from Two Experiments on Motion Sickness: Time (minutes) to Vomiting

Experiment 1		Experiment 2	
1	30	1	5
2	50	2	6*
3	50*	3	11
4	51	4	11
5	66*	5	13
6	82	6	24
7	92	7	63
8	120*	8	65
9	120*	9	69
10	120*	10	69
11	120*	11	79
12	120*	12	82
13	120*	13	82
14	120*	14	102
15	120*	15	115
16	120*	16	120*
17	120*	17	120*
18	120*	18	120*
19	120*	19	120*
20	120*	20	120*
21	120*	21	120*
		22	120*
		23	120*
		24	120*
		25	120*
		26	120*
		27	120*
		28	120*

Experiment 1: 0.167 Hz frequency, 0.11 g acceleration.
Experiment 2: 0.333 Hz frequency, 0.222 g acceleration.
* Indicates a censored observation.

1. Survival data are usually not symmetrically distributed — they will often be positively skewed, with a few people surviving a very long time compared with the majority; so assuming, for example, a normal distribution will not be reasonable.

Table 8.2 Bird Survival

Bird	Time	Indicator	Age	Weight	Length
1	2	1	1	1160	277
2	6	0	0	1140	266
3	6	0	1	1260	280
4	7	1	0	1160	264
5	13	1	1	1080	267
6	14	0	0	1120	262
7	16	0	1	1140	277
8	16	1	1	1200	283
9	17	0	1	1100	264
10	17	1	1	1420	270
11	20	0	1	1120	272
12	21	1	1	1110	271
13	22	1	0	1070	268
14	26	1	0	940	252
15	26	1	0	1240	271
16	27	1	0	1120	265
17	28	0	1	1340	275
18	29	1	0	1010	272
19	32	1	0	1040	270
20	32	0	1	1250	276
21	34	1	0	1200	276
22	34	1	0	1280	270
23	37	1	0	1250	272
24	40	1	0	1090	275
25	41	1	1	1050	275
26	44	1	0	1040	255
27	49	0	0	1130	268
28	54	0	1	1320	285
29	56	0	0	1180	259
30	56	0	0	1070	267
31	57	0	1	1260	269
32	57	0	0	1270	276
33	58	0	0	1080	260
34	63	0	1	1110	270
35	63	0	0	1150	271
36	63	0	0	1030	265
37	63	0	0	1160	275
38	63	0	0	1180	263
39	63	0	0	1050	271
40	63	0	1	1280	281

Table 8.2 (Continued) Bird Survival

Bird	Time	Indicator	Age	Weight	Length
41	63	0	0	1050	275
42	63	0	0	1160	266
43	63	0	0	1150	263
44	63	0	1	1270	270
45	63	0	1	1370	275
46	63	0	1	1220	265
47	63	0	0	1220	268
48	63	0	0	1140	262
49	63	0	0	1140	270
50	63	0	0	1120	274

Time: days observed.
Indicator: 0 censored observation, 1: observed death
Age 0: hatch-year birds, 1: after-hatch-year birds.
Weight: weight of bird (g).
Length: wing length of bird (mm).

2. At the completion of the study, some participants may not have reached the end point of interest. Consequently, their exact survival times are not known. All that is known is that the survival times are greater than the amount of time the individual has been in the study. The survival times of these participants are said to be *censored* (more precisely, they are *right-censored*).

8.2 Describing Survival Times and Cox's Regression

Of central importance in the analysis of survival data are two functions describing the distribution of survival times, the *survival function* and the *hazard function*.

8.2.1 The Survival Function

Using T to denote survival time, the survival function, $S(t)$, is defined as the probability that an individual survives longer than t,

$$S(t) = Pr(T > t) \tag{8.1}$$

The graph of $S(t)$ against t is known as the survival curve. When there are no censored observations the survivor function can be estimated very simply as

$$\hat{S}(t) = \frac{\text{number of individuals with survival times} \geq t}{\text{number of individuals in the data set}} \qquad (8.2)$$

But when the data contain censored observations a more complex estimator of $\hat{S}(t)$ is required, the one most commonly used is the *Kaplan–Meier estimator*, which is described in detail in Collett (1994).

For the data in Table 8.1 the main question of interest is whether the time to vomiting differs between the two experimental conditions. A graph of the estimated survival curves of each group will allow an informal assessment of this question; more formally, the survival experience of the two groups can be compared using what is known as the *log-rank test* (again, for details see Collett, 1994).

8.2.2 The Hazard Function

In the analysis of survival data, it is often of interest to assess which periods have high or low chances of the event of interest occurring amongst those alive at the time. Such risks can be quantified with the *hazard function*, defined as the probability that a participant experiences the event in a small time interval, s, given that the participant has survived up to the beginning of the interval, when the size of the time interval approaches zero, i.e.,

$$h(t) = \lim_{s \to 0} \Pr\left(t \leq T < t + s \mid T \geq t\right) \qquad (8.3)$$

The conditioning aspect is very important. For example, the probability of dying at age 100 is very small because most people die before that age, but the probability of a person dying at age 100 given that the person has reached that age is much greater. (The hazard function is also often known as the *instantaneous failure rate*.)

The hazard function is of most use as the basis for modelling the dependence of survival time on a number of explanatory variables (compare multiple regression, Chapter 4). Since $h(t)$ is restricted to being positive, its logarithm is expressed as a linear function of the explanatory variables, plus an unspecified function of time $\alpha(t)$:

$$\log h(t) = \log \alpha(t) + \beta_1 x_1 + \ldots \beta_p x_p \qquad (8.4)$$

where $x_1, \ldots, x_p$ are the explanatory variables of interest. This model was proposed by Cox (1972) and is known as either *Cox's regression* or the *proportional hazards model*. The latter name arises because for any two individuals at any point in time, the ratio of their hazard functions is a constant. Because the baseline hazard function $\alpha(t)$ does not have to be specified explicitly, the proportional hazards model is, essentially, non-parametric.

The estimated regression coefficients give the change in the log hazard produced by a change of one unit in the corresponding explanatory variable. Interpretation is aided by exponentiating the coefficients to give the effects in terms of the hazard function directly.

It is Cox's regression model we shall use to investigate the data in Table 8.2 to assess how survival time of the ducks is related to the other variables: weight, age, and length.

8.3 Analysis Using S-PLUS

S-PLUS has extensive features for the analysis of survival data, many of them described in detail in Therneau and Grambach (2000). Here we shall illustrate only the most commonly used of these.

8.3.1 Motion Sickness

The motion sickness data in Table 8.1 are available as the data frame **motion** the contents of which are shown in Table 8.3.

We begin by calculating the Kaplan–Meier estimates of the survival curves for the two experiments and plotting these curves. We will use the command language for this, in particular the S-PLUS functions **Surv** and **survfit**.

```
attach(motion)
motion.surv<-Surv(Time,Status)
#use surv to create an S-PLUS survival
#object for use in other survival analysis
#functions
survfit(motion.surv~Group)
#use survfit function to get information
#about survival time characteristics of each group
summary(survfit(motion.surv~Group))
#use summary to get details of the calculation of each
#group's survival curve
plot(survfit(motion.surv~Group),lty = 1:2)
legend(locator(1),c("Experiment 1","Experiment 2"),
lty = 1:2)
```

Table 8.3   Contents of motion Data Frame

< motion

	Subject	Group	Status	Time
1	1	0	1	30
2	2	0	1	50
3	3	0	0	50
4	4	0	1	51
5	5	0	0	66
6	6	0	1	82
7	7	0	1	92
8	8	0	0	120
9	9	0	0	120
10	10	0	0	120
11	11	0	0	120
12	12	0	0	120
13	13	0	0	120
14	14	0	0	120
15	15	0	0	120
16	16	0	0	120
17	17	0	0	120
18	18	0	0	120
19	19	0	0	120
20	20	0	0	120
21	21	0	0	120
22	22	1	1	5
23	23	1	0	6
24	24	1	1	11
25	25	1	1	11
26	26	1	1	13
27	27	1	1	24
28	28	1	1	63
29	29	1	1	65
30	30	1	1	69
31	31	1	1	69
32	32	1	1	79
33	33	1	1	82
34	34	1	1	82
35	35	1	1	102
36	36	1	1	115
37	37	1	0	120
38	38	1	0	120
39	39	1	0	120

Table 8.3 (continued) Contents of motion Data Frame

	Subject	Group	Status	Time
< motion				
40	40	1	0	120
41	41	1	0	120
42	42	1	0	120
43	43	1	0	120
44	44	1	0	120
45	45	1	0	120
46	46	1	0	120
47	47	1	0	120
48	48	1	0	120
49	49	1	0	120

```
#plot the two survival curves and add a
#suitable legend
```

The numerical results are shown in Table 8.4 and the plot in Figure 8.1. The plotted survival curves suggest perhaps that the time to vomiting is longer under the conditions of experiment 1 than experiment 2, although the large number of censored observations in each experiment makes this interpretation problematical.

A formal test of the difference between the survival curves of the two groups can be made using the log-rank test implemented using the **survdiff** function. This test operates by first computing the expected number of 'deaths' for each unique death time, or failure time in the data set, assuming that the chances of dying, given that participants are at risk, is the same in both groups. The total number of expected deaths is then calculated for each group by adding the expected number of deaths for each failure time. The test then compares the observed number of deaths in each group with the expected number of deaths via a chi-squared statistic.

```
survdiff(motion.surv~Group)
```

The results of applying the test are shown in Table 8.5. There is no strong evidence for a real difference in the survival distributions in the two experiments. Time to vomiting in each experiment appears to have the same distribution.

The log-rank test weights the contributions from all failure times equally, regardless of when they occurred. However, we have more precise information at the beginning when a larger proportion of participants are

Table 8.4 Characteristics of Motion Sickness Data and Details of Kaplan–Meier Estimation of Survival Curves

```
> motion.surv <- Surv(Time, Status)
> survfit(motion.surv ~ Group)
Call: survfit(formula = motion.surv ~ Group)
```

	n	events	mean	se(mean)	median	0.95LCL	0.95UCL
Group = 0	21	5	105.4	6.12	NA	NA	NA
Group = 1	28	14	87.1	7.80	115	79	NA

```
> summary(survfit(motion.surv ~ Group))
Call: survfit(formula = motion.surv ~ Group)
```

Group = 0

time	n.risk	n.event	survival	std.err	lower 95% CI	upper 95% CI
30	21	1	0.952	0.0465	0.866	1.000
50	20	1	0.905	0.0641	0.788	1.000
51	18	1	0.854	0.0778	0.715	1.000
82	16	1	0.801	0.0894	0.644	0.997
92	15	1	0.748	0.0981	0.578	0.967

Group = 1

time	n.risk	n.event	survival	std.err	lower 95% CI	upper 95% CI
5	28	1	0.964	0.0351	0.898	1.000
11	26	2	0.890	0.0599	0.780	1.000
13	24	1	0.853	0.0679	0.730	0.997
24	23	1	0.816	0.0744	0.682	0.976
63	22	1	0.779	0.0797	0.637	0.952
65	21	1	0.742	0.0841	0.594	0.926
69	20	2	0.668	0.0906	0.512	0.871
79	18	1	0.630	0.0928	0.472	0.841
82	17	2	0.556	0.0956	0.397	0.779
102	15	1	0.519	0.0961	0.361	0.746
115	14	1	0.482	0.0962	0.326	0.713

still 'alive'. Various tests have been proposed that give greater weight to earlier survival times. The *Peto–Wilcoxon* method, for example, weights observations by the Kaplan–Meier estimate of the proportion of participants still alive at each failure time. This method can be applied using the **survdiff** function but now including the argument rho = 1.

```
survidiff(motion.surv~Group, rho = 1)
```

The results of this test are also given in Table 8.5. They are very similar in this case to those from the log-rank test.

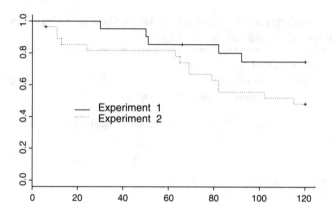

Figure 8.1 Survival curves for the data from each experiment in the motion sickness study.

Table 8.5 Results of Two Versions of the Log-Rank Test for the Motion Sickness Data

```
> survdiff(Surv(Time, Status) ~ Group)
Call:
survdiff(formula = Surv(Time, Status) ~ Group)
```

	N	Observed	Expected	(O-E)^2/E	(O-E)^2/V
Group = 0	21	5	8.86	1.68	3.21
Group = 1	28	14	10.14	1.47	3.21

```
   Chisq = 3.2 on 1 degrees of freedom, p = 0.0733
> survdiff(Surv(Time, Status) ~ Group, rho = 1)
Call:
survdiff(formula = Surv(Time, Status) ~ Group, rho = 1)
```

	N	Observed	Expected	(O-E)^2/E	(O-E)^2/V
Group = 0	21	4.01	7.2	1.41	3.22
Group = 1	28	11.49	8.3	1.22	3.22

Chisq = 3.2 on 1 degrees of freedom, p = 0.0728

8.3.2 Bird Deaths

The data in Table 8.2 are available as the data frame **bird**, the contents of which are given in Table 8.6.

Here we shall use the **Cox Proportional Hazards** dialog to fit a Cox's regression model to assess the effects of Age, Weight, and Length on a bird's survival time.

Table 8.6   Contents of bird Data Frame

> *bird*

	Bird	Time	Indicator	Age	Weight	Length
1	1	2	1	1	1160	277
2	2	6	0	0	1140	266
3	3	6	0	1	1260	280
4	4	7	1	0	1160	264
5	5	13	1	1	1080	267
6	6	14	0	0	1120	262
7	7	16	0	1	1140	277
8	8	16	1	1	1200	283
9	9	17	0	1	1100	264
10	10	17	1	1	1420	270
11	11	20	0	1	1120	272
12	12	21	1	1	1110	271
13	13	22	1	0	1070	268
14	14	26	1	0	940	252
15	15	26	1	0	1240	271
16	16	27	1	0	1120	265
17	17	28	0	1	1340	275
18	18	29	1	0	1010	272
19	19	32	1	0	1040	270
20	20	32	0	1	1250	276
21	21	34	1	0	1200	276
22	22	34	1	0	1280	270
23	23	37	1	0	1250	272
24	24	40	1	0	1090	275
25	25	41	1	1	1050	275
26	26	44	1	0	1040	255
27	27	49	0	0	1130	268
28	28	54	0	1	1320	285
29	29	56	0	0	1180	259
30	30	56	0	0	1070	267
31	31	57	0	1	1260	269
32	32	57	0	0	1270	276
33	33	58	0	0	1080	260
34	34	63	0	1	1110	270
35	35	63	0	0	1150	271
36	36	63	0	0	1030	265
37	37	63	0	0	1160	275
38	38	63	0	0	1180	263
39	39	63	0	0	1050	271

Table 8.6 (Continued)   Contents of bird Data Frame

> *bird*

	Bird	Time	Indicator	Age	Weight	Length
40	40	63	0	1	1280	281
41	41	63	0	0	1050	275
42	42	63	0	0	1160	266
43	43	63	0	0	1150	263
44	44	63	0	1	1270	270
45	45	63	0	1	1370	275
46	46	63	0	1	1220	265
47	47	63	0	0	1220	268
48	48	63	0	0	1140	262
49	49	63	0	0	1140	270
50	50	63	0	0	1120	274

- Click on **Statistics**.
- Select **Survival**.
- Select **Cox Proportional Hazards**.

The **Cox Proportional Hazards** dialog now appears. Select **bird** as the data set and then click on **Create Formula**. The **Formula** dialog now become available.

- Highlight **Time** and click on the **Time 1** tab.
- Highlight **Indicator** and click on the **Censor Codes** tab.
- Click on the **Add Response** tab.

This creates a suitable S-PLUS survival object as the response. Now highlight Age, Weight, and Length and click in the **Main Effect (+)** tab. The **Formula** dialog now appears as shown in Figure 8.2.

Click **OK** to recover the **Cox Proportional Hazards** dialog and again click **OK**. The results of fitting the model are given in Table 8.7. Here the tests of the individual regression coefficients suggest that none is significantly different from zero. This is confirmed by Wald's and the other two tests that assess the three coefficients simultaneously. The *p*-values associated with each of these indicate that here the three explanatory variables have no effect on the survival times of the ducks.

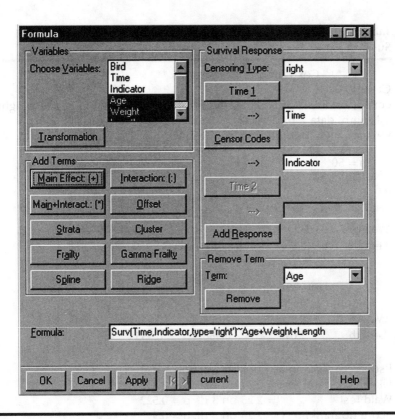

Figure 8.2 Completed Formula dialog in the Cox Proportional Hazards dialog.

Table 8.7   Results of Fitting Proportional Hazards Model to Bird Survival Data

```
            *** Cox Proportional Hazards ***
Call:
coxph(formula = Surv(Time, Indicator, type = "right") ~ Age + Weight
+ Length, data = bird, na.action = na.exclude, eps = 0.0001,
iter.max = 10, method = "efron", robust = F)

    n = 50
```

	coef	exp(coef)	se(coef)	z	p
Age	0.4637	1.590	0.57810	0.802	0.42
Weight	−0.0042	0.996	0.00289	−1.451	0.15
Length	0.0125	1.013	0.04087	0.306	0.76

	exp(coef)	exp(−coef)	lower .95	upper .95
Age	1.590	0.629	0.512	4.94
Weight	0.996	1.004	0.990	1.00
Length	1.013	0.988	0.935	1.10

```
Rsquare = 0.043 (max possible = 0.923)
Likelihood ratio test = 2.2 on 3 df,   p = 0.532
Wald test             = 2.25 on 3 df, p = 0.522
Score (logrank) test  = 2.28 on 3 df, p = 0.517
```

Exercises

8.1. Construct separate plots of the survival curves of the data from each experiment on motion sickness, showing in each case the estimated 95% confidence interval.

8.2. Plot the estimated survival curves of the hatch-year and after-hatch-year birds obtained from the **bird** data frame.

8.3. Fit a stratified Cox's model to the data in the **bird** data frame to explore the possibility that there should be separate baseline hazards for hatch-year and after-hatch-year birds.

Chapter 9

Exploring Multivariate Data: Male Egyptian Skulls

9.1 Description of Data

In this chapter we shall analyse the data shown in Table 9.1. These data show four measurements on male Egyptian skulls from five epochs. The measurements are

> **MB**: Maximum Breadth
> **BH**: Basibregmatic Height
> **BL**: Basialiveolar Length
> **NH**: Nasal Height

One question about these data is whether there is evidence of any change in the skulls over time? A steady change of head shape with time would indicate interbreeding with immigrant populations.

9.2 Exploring Multivariate Data

In this chapter we shall use a variety of primarily informal graphical methods to explore the structure of the data in Table 9.1 keeping in mind the main question of interest about the data. Some of these methods will have been used in previous chapters, but here we shall also use *principal components analysis* (see Everitt and Dunn, 2001) to help simplify the

Table 9.1 Data on Male Egyptian Skulls

c 4000 BC				c 3300 BC				c 1850 BC				c 200 BC				c AD 150			
MB	BH	BL	NH	MB	BH	BL	NH	MB	BH	BL	NH	MB	BH	BL	NH	MB	BH	BL	NH
131	138	89	49	124	138	101	48	137	141	96	52	137	134	107	54	137	123	91	50
125	131	92	48	133	134	97	48	129	133	92	47	141	128	95	53	136	131	95	49
131	132	99	50	138	134	98	45	132	138	87	48	141	130	87	49	128	126	91	57
119	132	96	44	148	129	104	51	130	134	106	50	135	131	99	51	130	134	92	52
136	143	100	54	126	124	95	45	134	134	96	45	133	120	91	46	138	127	86	47
138	137	89	56	135	136	98	52	140	133	98	50	131	135	90	50	126	138	101	52
139	130	108	48	132	145	100	54	138	138	95	47	140	137	94	60	136	138	97	58
125	136	93	48	133	130	102	48	136	145	99	55	139	130	90	48	126	126	92	45
131	134	102	51	131	134	96	50	136	131	92	46	140	134	90	51	132	132	99	55
134	134	99	51	133	125	94	46	126	136	95	56	138	140	100	52	139	135	92	54
129	138	95	50	133	136	103	53	137	129	100	53	132	133	90	53	143	120	95	51
134	121	95	53	131	139	98	51	137	139	97	50	134	134	97	54	141	136	101	54
126	129	109	51	131	136	99	56	136	126	101	50	135	135	99	50	135	135	95	50
132	136	100	50	138	134	98	49	137	133	90	49	133	136	95	52	137	134	93	53

141	140	100	51	130	136	104	53	129	142	104	47	136	130	99	55	142	135	96	52
131	134	97	54	131	128	98	45	135	138	102	55	134	137	93	542	139	134	95	47
135	137	103	50	138	129	107	53	129	135	92	50	131	141	99	55	138	125	99	51
132	133	93	53	123	131	101	51	134	125	90	60	129	135	95	47	137	135	96	54
139	136	96	50	130	129	105	47	138	134	96	51	136	128	93	54	133	125	92	50
132	131	101	49	134	130	93	54	136	135	94	53	131	125	88	48	145	129	89	47
126	133	102	51	137	136	106	49	132	130	91	52	139	130	94	53	138	136	92	46
135	135	103	47	126	131	100	48	133	131	100	50	144	124	86	50	131	129	97	44
134	124	93	53	135	136	97	52	138	137	94	51	141	131	97	53	143	126	88	54
128	134	103	50	129	126	91	50	130	127	99	45	130	131	98	53	131	124	91	55
130	130	104	49	134	139	101	49	136	133	91	49	133	128	92	51	132	127	97	52
138	135	100	55	131	134	90	53	134	123	95	52	138	126	97	54	137	125	85	57
128	132	93	53	132	130	104	50	136	137	101	54	131	142	95	53	129	128	81	52
127	129	106	48	130	132	93	52	133	131	96	49	136	138	94	55	140	135	103	48
131	136	114	54	135	132	98	54	138	133	100	55	132	136	92	52	147	129	87	48
124	138	101	46	130	128	101	51	138	133	91	46	135	130	100	51	136	133	97	51

data, and *multidimensional scaling* (see Everitt and Rabe-Hesketh, 1999) to display graphically *Mahalanobis distances* (see Chapter 2) between the five epochs.

Principal components analysis seeks linear compounds of the original variables, which are uncorrelated and which account for maximal amounts of the variation in these variables. By using these new variables it may be possible to achieve a parsimonious summary of the data in a reduced number of dimensions.

Multidimensional scaling attempts to find a low-dimensional representation of an observed distance matrix so that the Euclidean distances in this representation match as closely as possible in some sense the observed distances. In this way it may be possible to find an adequate graphical representation of the distances that enables any structure or pattern to be discovered.

9.3 Analysis Using S-PLUS

We assume that the data in Table 9.1 are available as an S-PLUS data frame **skulls**, which as well as the four measurements described in the previous section, contains a factor variable, EPOCH, with five levels labelling the epoch at which the measurements were made. The contents of **skulls** is shown in Table 9.2.

To begin we shall construct some simple, but nonetheless informative graphics using the two graphics palettes which we remind readers can be made visible by clicking

 (2D)

and

(3D)

We shall also make use of the 'conditioning' feature of S-PLUS that allows any type of graph to be plotted, conditional on the values of some variable (or variables). Here the conditioning will involve the categorical variable, EPOCH, but continuous variables may also be used, in which case the variable values are divided into a small number of intervals. This conditioning feature of S-PLUS has the potential to be very powerful, and we suggest readers experiment with the possibilities.

Table 9.2   Contents of **skulls** Data Frame

> *skulls*

	EPOCH	MB	BH	BL	NH
1	c4000BC	131	138	89	49
2	c4000BC	125	131	92	48
3	c4000BC	131	132	99	50
4	c4000BC	119	132	96	44
5	c4000BC	136	143	100	54
6	c4000BC	138	137	89	56
7	c4000BC	139	130	108	48
8	c4000BC	125	136	93	48
9	c4000BC	131	134	102	51
10	c4000BC	134	134	99	51
11	c4000BC	129	138	95	50
12	c4000BC	134	121	95	53
13	c4000BC	126	129	109	51
14	c4000BC	132	136	100	50
15	c4000BC	141	140	100	51
16	c4000BC	131	134	97	54
17	c4000BC	135	137	103	50
18	c4000BC	132	133	93	53
19	c4000BC	139	136	96	50
20	c4000BC	132	131	101	49
21	c4000BC	126	133	102	51
22	c4000BC	135	135	103	47
23	c4000BC	134	124	93	53
24	c4000BC	128	134	103	50
25	c4000BC	130	130	104	49
26	c4000BC	138	135	100	55
27	c4000BC	128	132	93	53
28	c4000BC	127	129	106	48
29	c4000BC	131	136	114	54
30	c4000BC	124	138	101	46
31	c3300BC	124	138	101	48
32	c3300BC	133	134	97	48
33	c3300BC	138	134	98	45
34	c3300BC	148	129	104	51
35	c3300BC	126	124	95	45
36	c3300BC	135	136	98	52
37	c3300BC	132	145	100	54
38	c3300BC	133	130	102	48
39	c3300BC	131	134	96	50

**Table 9.2 (Continued)   Contents of skulls
Data Frame**

> skulls

	EPOCH	MB	BH	BL	NH
40	c3300BC	133	125	94	46
41	c3300BC	133	136	103	53
42	c3300BC	131	139	98	51
43	c3300BC	131	136	99	56
44	c3300BC	138	134	98	49
45	c3300BC	130	136	104	53
46	c3300BC	131	128	98	45
47	c3300BC	138	129	107	53
48	c3300BC	123	131	101	51
49	c3300BC	130	129	105	47
50	c3300BC	134	130	93	54
51	c3300BC	137	136	106	49
52	c3300BC	126	131	100	48
53	c3300BC	135	136	97	52
54	c3300BC	129	126	91	50
55	c3300BC	134	139	101	49
56	c3300BC	131	134	90	53
57	c3300BC	132	130	104	50
58	c3300BC	130	132	93	52
59	c3300BC	135	132	98	54
60	c3300BC	130	128	101	51
61	c1850BC	137	141	96	52
62	c1850BC	129	133	93	47
63	c1850BC	132	138	87	48
64	c1850BC	130	134	106	50
65	c1850BC	134	134	96	45
66	c1850BC	140	133	98	50
67	c1850BC	138	138	95	47
68	c1850BC	136	145	99	55
69	c1850BC	136	131	92	46
70	c1850BC	126	136	95	56
71	c1850BC	137	129	100	53
72	c1850BC	137	139	97	50
73	c1850BC	136	126	101	50
74	c1850BC	137	133	90	49
75	c1850BC	129	142	104	47
76	c1850BC	135	138	102	55

Table 9.2 (Continued)   Contents of skulls Data Frame

> *skulls*

	EPOCH	MB	BH	BL	NH
77	c1850BC	129	135	92	50
78	c1850BC	134	125	90	60
79	c1850BC	138	134	96	51
80	c1850BC	136	135	94	53
81	c1850BC	132	130	91	52
82	c1850BC	133	131	100	50
83	c1850BC	138	137	94	51
84	c1850BC	130	127	99	45
85	c1850BC	136	133	91	49
86	c1850BC	134	123	95	52
87	c1850BC	136	137	101	54
88	c1850BC	133	131	96	49
89	c1850BC	138	133	100	55
90	c1850BC	138	133	91	46
91	c200BC	137	134	107	54
92	c200BC	141	128	95	53
93	c200BC	141	130	87	49
94	c200BC	135	131	99	51
95	c200BC	133	120	91	46
96	c200BC	131	135	90	50
97	c200BC	140	137	94	60
98	c200BC	139	130	90	48
99	c200BC	140	134	90	51
100	c200BC	138	140	100	52
101	c200BC	132	133	90	53
102	c200BC	134	134	97	54
103	c200BC	135	135	99	50
104	c200BC	133	136	95	52
105	c200BC	136	130	99	55
106	c200BC	134	137	93	52
107	c200BC	131	141	99	55
108	c200BC	129	135	95	47
109	c200BC	136	128	93	54
110	c200BC	131	125	88	48
111	c200BC	139	130	94	53
112	c200BC	144	124	86	50
113	c200BC	141	131	97	53

Table 9.2 (Continued)   Contents of skulls Data Frame

> *skulls*

	EPOCH	MB	BH	BL	NH
114	c200BC	130	131	98	53
115	c200BC	133	128	92	51
116	c200BC	138	126	97	54
117	c200BC	131	142	95	53
118	c200BC	136	138	94	55
119	c200BC	132	136	92	52
120	c200BC	135	130	100	51
121	cAD150	137	123	91	50
122	cAD150	136	131	95	49
123	cAD150	128	126	91	57
124	cAD150	130	134	92	52
125	cAD150	138	127	86	47
126	cAD150	126	138	101	52
127	cAD150	136	138	97	58
128	cAD150	126	126	92	45
129	cAD150	132	132	99	55
130	cAD150	139	135	92	54
131	cAD150	143	120	95	51
132	cAD150	141	136	101	54
133	cAD150	135	135	95	56
134	cAD150	137	134	93	53
135	cAD150	142	135	96	52
136	cAD150	139	134	95	47
137	cAD150	138	125	99	51
138	cAD150	137	135	96	54
139	cAD150	133	125	92	50
140	cAD150	145	129	89	47
141	cAD150	138	136	92	46
142	cAD150	131	129	97	44
143	cAD150	143	126	88	54
144	cAD150	134	124	91	55
145	cAD150	132	127	97	52
146	cAD150	137	125	85	57
147	cAD150	129	128	81	52
148	cAD150	140	135	103	48
149	cAD150	147	129	87	48
150	cAD150	136	133	97	51

	1	2	3	4	5	6	7	8	9	10
	EPOCH	MB	BH	BL	NH					
1	c4000BC	131.00	138.00	89.00	49.00					
2	c4000BC	125.00	131.00	92.00	48.00					
3	c4000BC	131.00	132.00	99.00	50.00					
4	c4000BC	119.00	132.00	96.00	44.00					
5	c4000BC	136.00	143.00	100.00	54.00					
6	c4000BC	138.00	137.00	89.00	56.00					
7	c4000BC	139.00	130.00	108.00	48.00					
8	c4000BC	125.00	136.00	93.00	48.00					
9	c4000BC	131.00	134.00	102.00	51.00					
10	c4000BC	134.00	134.00	99.00	51.00					
11	c4000BC	129.00	138.00	95.00	50.00					
12	c4000BC	134.00	121.00	95.00	53.00					
13	c4000BC	126.00	129.00	109.00	51.00					
14	c4000BC	132.00	136.00	100.00	50.00					
15	c4000BC	141.00	140.00	100.00	51.00					

Figure 9.1 Skulls data and 2D and 3D graphics.

To access the **skulls** data frame for use with the two graphics palettes:

- Click on **Data**.
- Choose **Select Data**.
- In the **Select Data** dialog that now appears choose **skulls** in the **Existing Data** box and then click **OK**.

The **skulls** data frame appears, and we now make the two graphics palettes available by clicking on the appropriate buttons. Also, we need to turn the condition mode on by clicking on

(this button "toggles" between conditioning on and off and the 1 refers to the number of conditioning variables, which can be altered). The screen now looks something like Figure 9.1.

To begin, suppose we require histograms of, say, MB for each epoch.

- Highlight the **MB** column by clicking on the name.
- Now **Ctrl click** the **EPOCH** column.
- Select the **Histogram** option on the **2D Graph palette**.

This leads to a coloured version of Figure 9.2. The first variable highlighted is the one that forms the histograms, the second is the conditioning variable. There is perhaps some evidence that MB increases over the epochs. Similar graphs could now be constructed for the other three measurements in the data set.

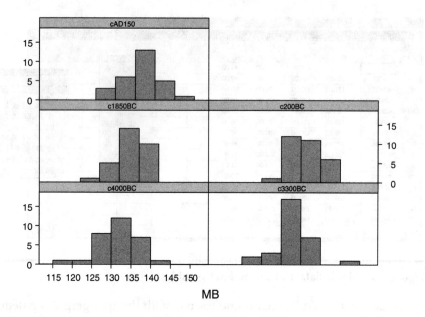

Figure 9.2 Histograms of the variable MB for each epoch.

Now suppose we would like a scatterplot of BH against MG for each epoch, with a locally weighted regression fit drawn on each graph.

- Highlight **MB** (this will be the *x* variable).
- **Ctrl click** on **BH** (this will be the *y* variable).
- **Ctrl click** on **EPOCH** (since the conditioning mode is on, this will be the conditioning variable).
- Select **Loess** on the **2D Graph palette**.

This series of operations leads to the graph in Figure 9.3. Is there a different relationship between MB and BH in the different epochs? It is perhaps difficult to judge, although there appears to be a small but definite negative relationship in c200BC that is not present in the other epochs.

Again, similar diagrams could be produced for all other pairs of variables. All such diagrams can be summarized conveniently by constructing a scatterplot matrix for each of the epochs.

- Highlight **MB** and the **Ctrl click** on each of the other three measurement variables.
- **Ctrl click** on **EPOCH**.
- Select the **Scatter Matrix** option in the **2D Graph palette**.

This leads to Figure 9.4. Although this is a rather 'busy' diagram, it does enable all the pairwise scatterplots for the five different epochs to be

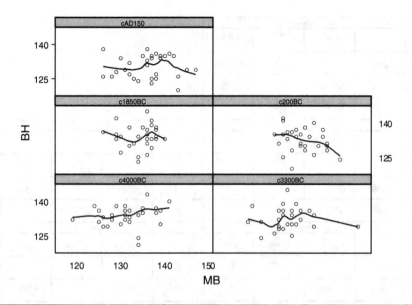

Figure 9.3   Scatterplots of the variables MB and BH for each epoch, showing a locally weighted regression fit.

viewed together, which might be helpful to an experienced archaeologist if not to the author!

Now let us consider two examples of using the **3D graph palette**, very much as exercises in the use of this feature of S-PLUS, rather than particularly helpful graphics for exploring the skulls data. To get two, three-dimensional graphs of the variables MB, BH, and BL (here we do not want the plots to be conditioned on epoch and so the conditioning button must be in the off position):

- Highlight **MB**.
- **Ctrl click** on **BH** and on **BL**.
- Select **3D Scatter** in the **3D Graph palette**.

This gives Figure 9.5. Such diagrams are often made clearer by 'dropping' lines from the plotted points to a plane through the origin. To construct this diagram, proceed as for Figure 9.5 but Select **Drop Line Scatter** in the **3D Graph palette** to get Figure 9.6.

Now let us consider some more complex methods for exploring the structure of the skulls data. To begin, we shall apply principal components to the data. This technique, described fully in Everitt and Dunn (2001) attempts to reduce the dimensionality of multivariate data by finding uncorrelated linear combinations of the original variables that account for maximal proportions of the variation in the data. Such variables can be

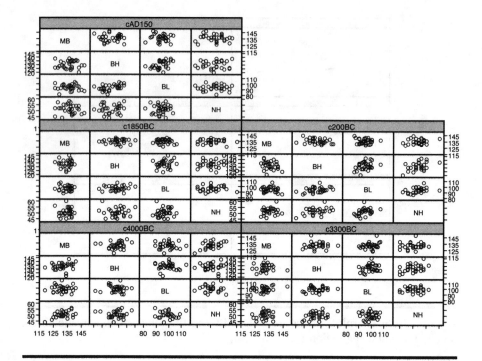

Figure 9.4   Scatterplot matrix of variables in the skulls data frame for each epoch.

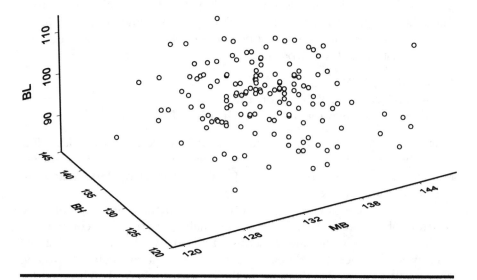

Figure 9.5   Three-dimensional scatter of the variables BH, MB, and BL.

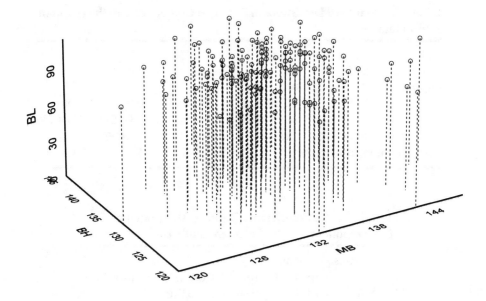

Figure 9.6   Three-dimensional drop-line scatter of the three variables BH, MB, and BL.

used in a variety of ways, but here our main interest will be in using the derived components to obtain an informative graphical display of the data. For this analysis we shall use the command language approach.

```
skulls.pc<-princomp(skulls[,-1],cor=T)
#produces a principal components of the skulls
#data after removing EPOCH. Components are
#derived from the correlation matrix
summary(skulls.pc)
```

This gives the results shown in Table 9.3. The first component accounts for just over 33% of the variance; the first two for 64%.

The loadings of the four observed variables on each component are displayed using the command:

```
skulls.pc$loadings
```

The results appear in Table 9.4.

The first component is a contrast between MB on the one hand and BH and BL on the other. The second component is simply a weighted average of MB, BH, and NH. Interpretation of the components is not our

Table 9.3   Standard Deviations of Principal Components of the Egyptian Skulls Data

```
> skulls.pc <- princomp(skulls[, -1], cor = T)
> summary(skulls.pc)
Importance of components:
```

	Comp. 1	Comp. 2	Comp. 3	Comp. 4
Standard deviation	1.1564147	1.0983595	0.8731551	0.8330136
Proportion of Variance	0.3343237	0.3015984	0.1906000	0.1734779
Cumulative Proportion	0.3343237	0.6359221	0.8265221	1.0000000

Table 9.4   Coefficients Defining the Principal Components of the Egyptian Skulls Data

```
> skulls.pc$loadings
```

	Comp. 1	Comp. 2	Comp. 3	Comp. 4
MB	−0.407	0.567	−0.710	
BH	0.617	0.345		−0.707
BL	0.672		−0.469	0.572
NH		0.748	0.525	0.405

main aim here, however; rather we shall use the scores of each skull on the first two components to display the data graphically.

```
skulls.pcx<-skull.pc$scores[,1]
skulls.pcy<-skull.pc$scores[,2]
#skulls.pcx and skulls.pcy now contain the
#first and second principal components scores
#for each skull
par(pty="s")
#select a square plotting area
xlim<-range(skulls.pcx)
#find range of x values to use in plotting
#the data
plot(skulls.pcx,skulls.pcy,xlim=xlim,
ylim=xlim,xlab="PC1",ylab="PC2",
type="n")
#set up axes of plot and label axis but do
#not plot the data
```

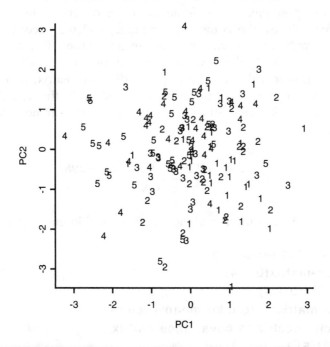

Figure 9.7 Plot of skulls data in the space of the first two principal component scores, with points labelled by epoch.

```
labs<-rep(1:5,rep(30,5))
#labs contains values 1 to 5 in sets of 30 and
#will be used to label the plot according to epoch
#since the EPOCH labels themselves would make
#the plot too messy
text(skulls.pcx,skulls.pcy,labels=labs)
```

The resulting diagram is shown in Figure 9.7. There does appear to be some distinction between epochs 1 (c4000BC) and 5(cAD150) in this diagram, but the observations for the remaining three epochs overlap to a considerable extent.

Finally, we shall calculate Mahalanobis distances (see Chapter 2) between each pair of epochs, and then try to represent the distances by Euclidean distances in two dimensions using *classical multidimensional scaling*. This technique is described in detail in Everitt and Rabe-Hesketh (1997), but in essence it tries to represent, in low-dimensional space, an observed distance matrix, by deriving coordinates for each observation (epochs in this case) so that the Euclidean distances defined by these coordinates approximate the elements of the observed distance matrix. Again, we shall use the command language, and we need in particular to use the **apply** function to obtain the mean vectors of each epoch, the **var** function to obtain the covariance matrix of each epoch, the **Mahalanobis** function to calculate the required distance matrix, and the **cmdscale** function to apply multidimensional scaling. In this calculation we shall use the following estimate of the assumed common covariance matrix, **S**,

$$\mathbf{S} = \frac{29\mathbf{S}_1 + 29\mathbf{S}_2 + 29\mathbf{S}_3 + 29\mathbf{S}_4 + 29\mathbf{S}_5}{145} \tag{9.1}$$

where $\mathbf{S}_1, \ldots, \mathbf{S}_5$, are the covariance matrices within each epoch.

```
#Mahalanobis distances
centres<-matrix(0,5,4)
S <-matrix(0,4,4)
#set up matrices to take mean vectors
#of each epoch and covariance matrix
for(i in 1:5) {
      centres[i,]<-apply(skulls[labs==i,-1],2,mean)
      S<-S+29*var(skulls[,-1])
}
#use for loop to calculate all mean vectors and the
#combined covariance matrix
S<-S/145
mahal<-matrix(0,5,5)
for(i in 1:5) {
      mahal[i,]<-mahalanobis(centres,centres[i,],S)
}
#calculate required matrix of Mahalanobis distances
#between epoch means
par(pty="s")
coords<-cmdscale(mahal)
xlim<-range(coords[, 1])
plot(coords,xlab="C1",ylab="C2",type="n",xlim=xlim,
ylim=xlim)
text(coords,labels=c("c4000BC","c3300BC","c1850BC",
"c200BC","cAD150"))
```

The Mahalanobis distance matrix is shown in Table 9.5. The resulting diagram is shown in Figure 9.8. Here there is a very striking confirmation of a steady change of head shape with time.

Table 9.5 Mahalanobis Distances between Epochs

	[,1]	[,2]	[,3]	[,4]	[,5]
[1,]	0.00000000	0.08544641	0.7040863	1.3632193	1.9644964
[2,]	0.08544641	0.00000000	0.5923855	1.1838453	1.5724978
[3,]	0.70408633	0.59238550	0.0000000	0.3679117	0.7388707
[4,]	1.36321930	1.18384525	0.3679117	0.0000000	0.2009955
[5,]	1.96449636	1.57249782	0.7388707	0.2009955	0.0000000

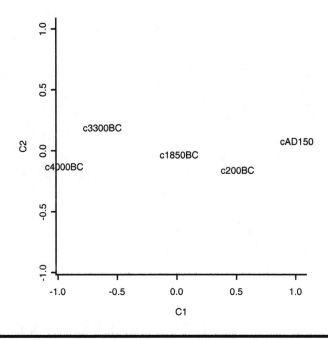

Figure 9.8 Two-dimensional solution from classical multidimensional scaling applied to the Mahalanobis distances between epochs.

Exercises

9.1. Construct a scatterplot matrix of all the principal component scores of the skulls data with the points in each panel labelled in some way by epoch.

9.2. Use the **Select Data** and **2D Graph palette** options to construct a contour plot of the bivariate density of the first two principal component scores of the skulls data.

9.3. Plot the convex hull of the observations of each epoch onto the plot of the data in the space of the first two principal components. (This will involve finding out about and then using the following S-PLUS functions: **chull** and **polygon**.)

Chapter 10

Cluster Analysis: Low Temperatures and Voting in Congress

10.1 Description of Data

Two sets of data will be of concern in this chapter. The first, shown in Table 10.1, gives the lowest temperatures (in °F) recorded for various cities in the United States. The second, in Table 10.2, records the number of times 15 congresspeople from New Jersey voted differently in the House of Representatives on 19 environmental bills.

One question of interest about the lowest temperatures data is whether the temperatures can be used to group the cities into distinct classes? And classification is also of interest for the voting data as an aid to see if the data imply party affiliations in the voting records of the 15 congresspeople.

10.2 Cluster Analysis

Cluster analysis is a generic term for a large number of relatively disparate techniques that seek to determine whether or not a data set contains distinct groups or clusters of observations and, if so, to find these groups. A detailed review of the area is given in Everitt et al. (2001).

Table 10.1   Lowest Temperatures in a Number of Cities in the United States

January	April	July	October
−8	26	53	28
−7	20	51	25
−44	−12	35	5
−12	16	54	28
−27	7	40	17
4	30	59	29
−25	−2	43	3
−8	23	57	25
53	57	67	64
12	31	62	33
−22	6	36	11
23	39	49	41
30	46	69	51
−17	23	51	26
−6	12	52	28
−23	5	44	13
17	32	61	34
−26	8	40	15
−16	13	33	8
24	31	43	34
0	29	43	28
−5	24	55	29

The most commonly used subset of clustering methods is that generally referred to as *agglomerative hierarchical methods.* These produce a series of hierarchical partitions of the observations, beginning at the stage of n separate single-member 'groups' (where n is the sample size), and ending with a single group containing all the observations. The series of steps is usually summarized in a tree-like diagram known as a *dendrogram.* In most applications, users will be interested in selecting one or two partitions from the series that in some sense best describe the data. A number of informal indicators of number of groups when using this type of clustering procedure are described in Everitt et al. (2001).

Most agglomerative hierarchical clustering methods operate on an $n \times n$ *distance* or *dissimilarity matrix* for the objects being clustered. In general, this will be calculated from the multivariate observations available for each object, and the most commonly used distance measure is *Euclidean* defined as

Table 10.2   Voting in Congress

Name (Party)	1	2	3	4	5	6	7	8	9	10	11	12	13	14	15
1 Hunt (R)	0	8	15	15	10	9	7	15	16	14	15	16	7	11	13
2 Sandman (R)	8	0	17	12	13	13	12	16	17	15	16	17	13	12	16
3 Howard (D)	15	17	0	9	16	12	15	5	5	6	5	4	11	10	7
4 Thompson (D)	15	12	9	0	14	12	13	10	8	8	8	6	15	10	7
5 Freylinghuysen (R)	10	13	16	14	0	8	9	13	14	12	12	12	10	11	11
6 Forsythe (R)	9	13	12	12	8	0	7	12	11	10	9	10	6	6	10
7 Widnall (R)	7	12	15	13	9	7	0	17	16	15	14	15	10	11	13
8 Roe (D)	15	16	5	10	13	12	17	0	4	5	5	3	12	7	6
9 Heltoski (D)	16	17	5	8	14	11	16	4	0	3	2	1	13	7	5
10 Rodino (D)	14	15	6	8	12	10	15	5	3	0	1	2	11	4	6
11 Minish (D)	15	16	5	8	12	9	14	5	2	1	0	1	12	5	5
12 Rinaldo (R)	16	17	4	6	12	10	15	3	1	2	1	0	12	6	4
13 Maraziti (R)	7	13	11	15	10	6	10	12	13	11	12	12	0	9	13
14 Daniels (D)	11	12	10	10	11	6	11	7	7	4	5	6	9	0	9
15 Patten (D)	13	16	7	7	11	10	13	6	5	6	5	4	13	9	0

$$d_{ij} = \sqrt{\sum_{k=1}^{p} \left(x_{ik} - x_{jk} \right)^2} \qquad (10.1)$$

where x_{ik} and x_{jk} represent the value of the kth variable for objects i and j, respectively. The number of variables is p. Occasionally, however, the dissimilarity matrix, which is to be the basis of clustering, can arise directly, as with the voting data described previously.

Differences between agglomerative hierarchical techniques occur because of the variety of ways in which distance between a group and a single object, or between two groups, can be defined. Three methods that are commonly used are

- Distance between two groups taken as the distance between their two closest members — results in *single linkage clustering*.
- Distance between two groups taken as the distance between their two most remote members — results in *complete linkage clustering*.
- Distance between two groups taken as the distance between the average of the intergroup pairwise distances — results in *average linkage*.

All these methods and their properties are described in Everitt et al. (2001).

Another frequently used type of cluster analysis is based on finding a partition of the observations into a particular number of groups so as to optimise some numerical measure, small (or large for some measures) values of which indicate a 'good' clustering. One such method is *k-means* which, given an initial partition into the required number of groups, attempts to find an improved solution by searching for the partition that minimises the total within-group sum of squares of the observations. Again, there are various possibilities for deciding on the 'best' number of groups for a data set — see Everitt et al. (2001) for details. One simple method is to plot the within-clusters sum of squares against number of clusters. This sum of squares decreases monotonically as the number of clusters increases, but in many cases the decrease flattens markedly from some value onwards. The location of such an 'elbow' in the plot is often taken to indicate the appropriate number of clusters. Tibshirani et al. (2001) attempt to make this approach to identifying the number of clusters more formal with the introduction of the so-called 'gap' statistic, which for a sample of size n and a particular number of clusters k is defined as

$$\mathrm{Gap}_n\big(k\big) = E_n^*\big\{\log\big(W_k\big)\big\} - \log\big(W_k\big) \qquad (10.2)$$

where W_k is the within-clusters sum of squares, and E_n^* denotes expectation from a reference distribution, for which Tibshirani et al. (2001) make two suggestions;

1. A uniform distribution in which each variable is generated from a uniform distribution over the range of the observed values for that variable.
2. A uniform distribution over a box aligned with the principal components of the data (see Tibshirani et al. (2001) for more details).

$E_n^*\{\log(W_k)\}$ is estimated by the average of B copies $\log(W_k^*)$ found by drawing a number of samples from the reference distribution, and the number of clusters is chosen using the following: $\hat{k}$ = smallest k such that Gap $(k) \geq$ Gap $(k + 1) - s_{k+1}$, where $s_k = \sqrt{(1 + 1/B)}\,\mathrm{SD}(k)$ and $\mathrm{SD}(k)$ is the standard deviation of the B replicates of log (W_k^*). Further details are given in Tibshirani et al. (2001).

This approach will be explored in Section 10.3.1.

10.3   Analysis Using S-PLUS

In the following two subsections we will describe the application of *k*-means clustering to the temperature data on U.S. cities, and then the application of a number of hierarchical clustering procedures to the voting in Congress data.

10.3.1 Clustering Cities in the United States on the Basis of their Year-Round Lowest Temperature

The data in Table 10.1 are available as an S-PLUS data frame **lowest**, the contents of which are shown in Table 10.3.

We shall apply the *k*-means method to these data using the appropriate dialog, which can be accessed as follows:

■ Click on **Statistics**.
■ Select **Cluster Analysis**.
■ Select **K-Means**.

Table 10.3   Contents of the lowest Data Frame

```
> lowest
```

	Cities	January	April	July	October
1	Atlanta	−8	26	53	28
2	Baltimore	−7	20	51	25
3	Bismark	−44	−12	35	5
4	Boston	−12	16	54	28
5	Chicago	−27	7	40	17
6	Dallas	4	30	59	29
7	Denver	−25	−2	43	3
8	El Paso	−8	23	57	25
9	Honolulu	53	57	67	64
10	Houston	12	31	62	33
11	Juneau	−22	6	36	11
12	LA	23	39	49	41
13	Miami	30	46	69	51
14	Nashville	−17	23	51	26
15	NY	−6	12	52	28
16	Omaha	−23	5	44	13
17	Phoenix	17	32	61	34
18	Portland	−26	8	40	15
19	Reno	−16	13	33	8
20	SF	24	31	43	34
21	Seattle	0	29	43	28
22	Washington	−5	24	55	29

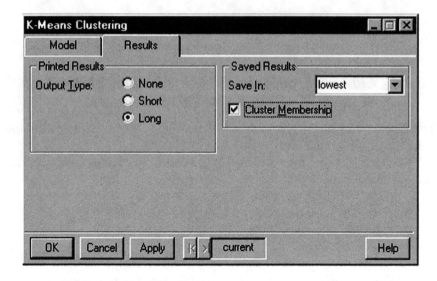

Figure 10.1   Part of *K*-Means Clustering dialog.

In the *k*-means clustering dialog that appears, select **lowest** as the data set and January, April, July, and October as the variables. Set the **Num of Clusters** to 3 and the **Max Iterations** to 50. Now click on the **Results** tab, choose **Long** for **Output Type**, check the **Cluster Membership** box, and save the cluster labels in the **lowest** data frame by selecting this data frame in the **Save In** slot. The dialog now looks as shown in Figure 10.1.

Click **OK** and the screen now looks like that shown in Figure 10.2 with the Report file giving the results of the cluster analysis and the **lowest** data frame now including the **cluster.id** variable both visible.

The results are shown in Table 10.4. The three clusters correspond to the following cities:

> Cluster 1: Atlanta, Baltimore, Boston, Dallas, El Paso, Nashville, New York, Seattle, Washington, D.C.
> Cluster 2: Bismarck, Chicago, Denver, Juneau, Omaha, Portland, Reno
> Cluster 3: Honolulu, Houston, Los Angeles, Miami, Phoenix, San Francisco

The cities in the second cluster are definitely to be avoided by those people fond of warm weather! Cluster 3 provides cities with much more appeal for such people. The cities in cluster 1 are intermediate in temperature between the other two clusters.

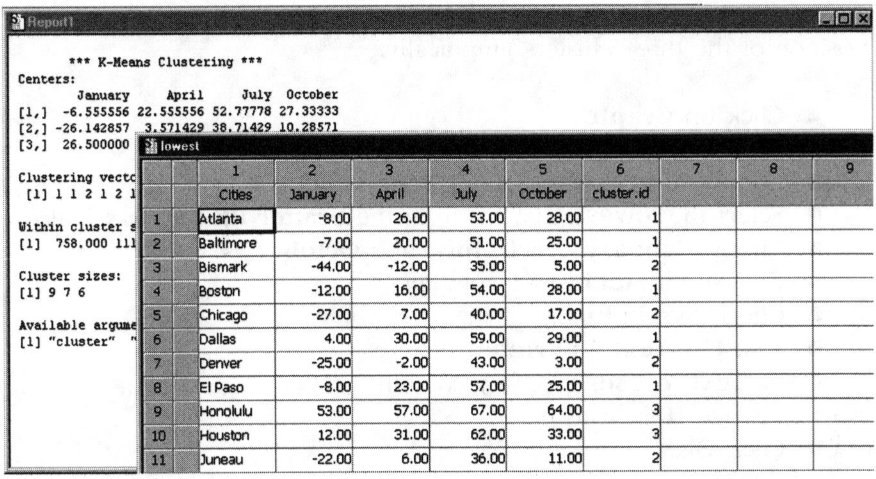

Figure 10.2 **Screen showing the revised lowest data frame now including cluster labels.**

Table 10.4 Results of *K*-Means Clustering on Lowest Temperature Data

*** K-Means Clustering ***

Centers:

	January	April	July	October
[1,]	–6.555556	22.555556	52.77778	27.33333
[2,]	–26.142857	3.571429	38.71429	10.28571
[3,]	26.500000	39.333333	58.50000	42.83333

Clustering vector:
[1] 1 1 2 1 2 1 2 1 3 3 2 3 3 1 1 2 3 2 2 3 1 1

Within cluster sum of squares:
[1]   758.000   1117.429   2885.167

Cluster sizes:
[1]   9   7   6

Available arguments:
[1]   "cluster"   "centers"   "withinss"   "size"

We can use the scatterplot matrix display to enable us to see the position of the three clusters graphically.

- Click on **Graph**.
- Select **2D Plot**.
- Choose **Matrix** in the **Axes Type** slot.
- Select the **lowest** data frame in the **Scatter Plot Matrix** dialog.
- Choose **January**, **April**, **July**, and **October** as *x* columns.
- Choose **cluster.id** as *y* columns.
- Check **Symbol** tab.
- Tick **Use Text As Symbol** box.
- As **Text to Use**, specify *y* column.
- Change **Height** to 0.15.
- Click **OK**.

The resulting plot is shown in Figure 10.3. The plot shows clearly the division into low-, medium-, and high-temperature cities.

The value of three for the number of clusters was chosen arbitrarily. It would be useful if the data themselves could provide some information

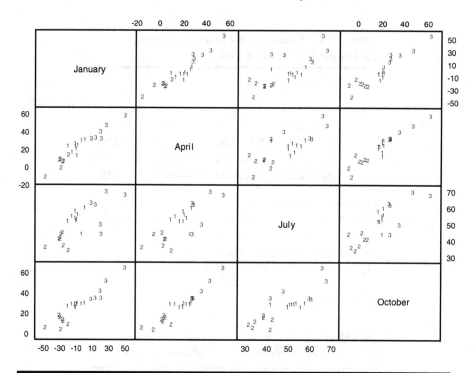

Figure 10.3   Scatterplot matrix of low-temperature data showing cluster labels.

regarding the likely number of clusters that best describes the data. To investigate this we shall use the suggestion made by Tibshirani et al. (2001), described in Section 10.2. A probably very inefficient S-PLUS function for computing the 'gap' statistic is as follows:

```
gap<-function(x,B=100,K=8,method="uniform") {
#
#K is the upper limit of number of clusters
#B is the number of samples from the reference distribution
#method="pc" if principal components approach wanted.
#
#get within cluster sums of squares for observed data
#for the requested upper limit of number of clusters
#
n<-length(x[,1])
p<-length(x[1,])
S<-var(x)
wss1<-(n-1)*sum(diag(S))
wsso<-numeric(0)
for(i in 2:K) {W<-sum(kmeans(x,i)$withinss)
               wsso<-c(wsso,W)
               }
wsso<-c(wss1,wsso)
#
if(method=="pc") {X <- sweep(x,2,apply(x,2,mean))
               X.svd<-svd(X)
               Xprime<-X%*%X.svd$v
                                  }
#
#get range of values for generating samples from the
reference distribution
#
if(method=="pc") rang<-apply(Xprime,2,range)
     else rang<-apply(x,2,range)
#
xstar<-matrix(0,n,p)
wk.exp<-matrix(0,B,K)
#
#now generate B samples from the reference distribution
```

```
#and cluster each, calculating the within cluster sum of
#squares for each solution
#
for(j in 1:B){
for(i in 1:n) xstar[i,]<-runif(p,rang[1,],rang[2,])
       if(method=="pc") xstar<-xstar%*%t(X.svd$v)
wss1<-(n-1)*sum(apply(xstar,2,var))
wss<-numeric(0)
for(i in 2:K) {W<-sum(kmeans(xstar,i)$withinss)
                 wss<-c(wss,W)
                 }
wss<-c(wss1,wss)
#
wk.exp[j,]<-wss
}
#
term<-apply(log(wk.exp),2,mean)
gap<-term-log(wsso)
#
sdk<-sqrt(apply(log(wk.exp),2,var))
#
sk<-sqrt(1+1/B)*sdk
#
#
                     diff<-gap[2:K]-sk[2:K]
                     diff<-gap[1:(K-1)]-diff
                  values<-seq(1:(K-1))[diff>=0]
                  nclust.est<-min(values)
results<-
list(gap=gap,sk=sk,lwss=log(wsso),lwkexp=term,B=B,K=K,
nclust.est=nclust.est)
results
}
```

A function to produce two useful plots from the object resulting from the **gap** function is as follows:

```
plotgap<-function(gapobj) {
    par(mfrow=c(1,2))
    ylim<-range(gapobj$lwss,gapobj$lwkexp)
```

```
plot(1:gapobj$K,gapobj$lwss,ylim=ylim,
ylab="Observed and expected log(wss)",
xlab="Number of clusters",type="l")
lines(1:gapobj$K,gapobj$lwkexp,lty=2)
legend(locator(1),c("Observed","Expected"),lty=1:2)
plot(1:gapobj$K,gapobj$gap,
xlab="Number of clusters",ylab="Gap statistic",type="l")
}
```

We can now apply these two functions to the low-temperature data:

```
#store numerical part of lowest data frame as a matrix
#
low<-as.matrix(lowest[,-1])
#apply gap statistic using each type of reference
    distribution in turn
set.seed(9251)
gap.res<-gap(low,B=50,K=6)
gap.res
plotgap(gap.res)
#
gap.res1<-gap(low,B=50,K=6,method="pc")
gap.res1
plotgap(gap.res1)
```

The results are shown in Table 10.5 and the plots in Figures 10.4 and 10.5. We see that the 'uniform' method estimates the number of clusters to be four, but the 'pc' method indicates that the observations are best regarded as a single homogeneous set. A detailed investigation of the gap statistic is given in Everitt (2002).

10.3.2 Classifying New Jersey Congresspeople on the Basis of their Voting Behaviour

We assume that the voting behaviour data are available as an S-PLUS matrix object **congress** and the names and party affiliations of the congresspeople as a vector **names**; see Table 10.6.

Here we shall use the command language approach to apply the three agglomerative hierarchical clustering methods mentioned in Section 10.2

Table 10.5 Results from Applying the gap Function to the Low-Temperature Data Using Both the 'Uniform' and the 'pc' Approach

```
> set.seed(9251)
> gap.res <- gap(low, B = 50, K = 6)
> gap.res
$gap:
[1]  0.378  0.629  1.113  1.401  1.458  1.466

$sk:
[1]  0.1075  0.0979  0.1115  0.1305  0.1421  0.1385

$lwss:
[1]  10.04  9.28  8.47  7.91  7.61  7.38

$lwkexp:
[1]  10.41  9.91  9.58  9.31  9.07  8.85

$B:
[1]  50

$K:
[1]  6

$nclust.est:
[1]  4

> gap.res1 <- gap(low, B = 50, K = 6, method = "pc")
> gap.res1
$gap:
[1]  0.3336  -0.1611  -0.0519  0.1055  0.1146  0.1082

$sk:
[1]  0.201  0.176  0.165  0.132  0.136  0.161

$lwss:
[1]  10.04  9.28  8.47  7.91  7.61  7.38

$lwkexp:
[1]  10.37  9.12  8.42  8.02  7.73  7.49

$B:
[1]  50

$K:
[1]  6

$nclust.est:
[1]  1
```

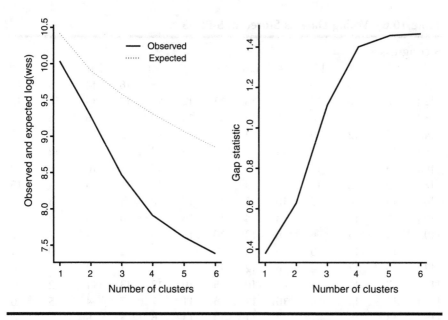

Figure 10.4   Plot of observed and expected log(wss) and gap statistic for low-temperature data using a uniform reference distribution.

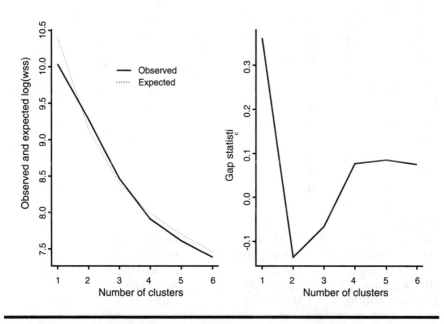

Figure 10.5   Plot of observed and expected log(wss) and gap statistic for low-temperature data using principal components reference distribution.

Table 10.6 Voting Data as Stored in S-PLUS

```
> congress
```

	[,1]	[,2]	[,3]	[,4]	[,5]	[,6]	[,7]	[,8]	[,9]	[,10]	[,11]	[,12]
[1,]	0	8	15	15	10	9	7	15	16	14	15	16
[2,]	8	0	17	12	13	13	12	16	17	15	16	17
[3,]	15	17	0	9	16	12	15	5	5	6	5	4
[4,]	15	12	9	0	14	12	13	10	8	8	8	6
[5,]	10	13	16	14	0	8	9	13	14	12	12	12
[6,]	9	13	12	12	8	0	7	12	11	10	9	10
[7,]	7	12	15	13	9	7	0	17	16	15	14	15
[8,]	15	16	5	10	13	12	17	0	4	5	5	3
[9,]	16	17	5	8	14	11	16	4	0	3	2	1
[10,]	14	15	6	8	12	10	15	5	3	0	1	2
[11,]	15	16	5	8	12	9	14	5	2	1	0	1
[12,]	16	17	4	6	12	10	15	3	1	2	1	0
[13,]	7	13	11	15	10	6	10	12	13	11	12	12
[14,]	11	12	10	10	11	6	11	7	7	4	5	6
[15,]	13	16	7	7	11	10	13	6	5	6	5	4

	[,13]	[,14]	[,15]
[1,]	7	11	13
[2,]	13	12	16
[3,]	11	10	7
[4,]	15	10	7
[5,]	10	11	11
[6,]	6	6	10
[7,]	10	11	13
[8,]	12	7	6
[9,]	13	7	5
[10,]	11	4	6
[11,]	12	5	5
[12,]	12	6	4
[13,]	0	9	13
[14,]	9	0	9
[15,]	13	9	0

```
> names
 [1] "Hunt(R)"        "Sandman(R)"          "Howard(D)"
 [4] "Thompson(D)"    "Freylinghuysen(R)"   "Forsythe(R)"
 [7] "Widnall(R)"     "Roe(D)"              "Heltoski(D)"
[10] "Rodino(D)"      "Minish(D)"           "Rinaldo(R)"
[13] "Maraziti(R)"    "Daniels(D)"          "Patten(D)"
```

to the data and to produce the resulting dendrograms. The two S-PLUS functions needed are **hclust** and **plclust**.

```
par(mfrow=c(1,3))
plclust(hclust(congress,method="connected"),
labels=names)
plclust(hclust(congress,method="compact"),
labels=names)
plclust(hclust(congress,method="average"),
labels=names)
#hclust takes a distance or dissimilarity matrix
#and applies either single linkage, method="connected",
#complete linkage, method="compact" or
#average linkage, method "average".
#plclust takes the result of hclust and produces
#the corresponding dendrogram, labelled
#here with the names of the congressmen
```

The resulting diagram showing the three dendrograms is given in Figure 10.6.

There are similarities but also differences in the three dendrograms, a situation that is not uncommon when applying different methods of cluster analysis to the same data set! Since single linkage is well known to have some practical problems (see Everitt et al., 2001), we shall confine more detailed attention to the results from complete and average linkage. Both dendrograms seem most consistent with a two-group solution, and the names of the congresspeople in each group can easily be extracted from Figure 10.6. But in applications with a larger number of observations, groups may be found using the **cuttree** function.

```
g<-cuttree(hclus(congress,method="compact"),k=2)
#stores labels for 2 cluster solution from
#complete linkage in g
names[g==1]
names[g==2]
#gives names of congressmen in two group
#solution
```

By simply changing to **method="average"** in the above, the names of the congresspeople in the two-group solution from average linkage are obtained. Both solutions are shown in Table 10.7. The two solutions are

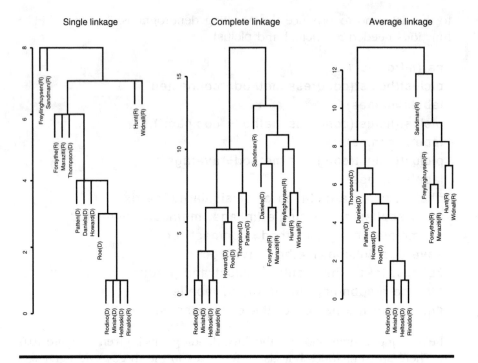

Figure 10.6   Dendrograms for single linkage, complete linkage, and average linkage for voting behaviour data.

very similar with only Daniels being placed in different clusters by the two clustering procedures. Both solutions show very definite evidence of party applications in voting.

The clustering results for these data can usefully be compared with the two-dimensional solution obtained by applying classical multidimensional scaling to the voting matrix

```
par(pty="s")
cords<-cmdscale(congress)
#gets 2D classical MDS solution
xlim<-range(cords[,1])
plot(cords,xlab="c1",ylab="c2",type="n",
xlim=xlim,ylim=xlim)
text(cords,labels=abbreviate(names))
#uses the abbreviate function to get unique
#shorter labels for each Congressman.
```

Table 10.7   Two Cluster Solutions from Complete Linkage and Average Linkage for the Voting in Congress Data

```
> g <- cutree(hclust(congress, method = "compact"), k = 2)
> names[g==1]
[1]   "Hunt(R)"              "Sandman(R)"          "Freylinghuysen(R)"
[4]   "Forsythe(R)"          "Widnall(R)"          "Maraziti(R)"
[7]   "Daniels(D)"
> names[g==2]
[1]   "Howard(D)"   "Thompson(D)"    "Roe(D)"        "Heltoski(D)"
[5]   "Rodino(D)"   "Minish(D)"      "Rinaldo(R)"    "Patten(D)"
> g <- cutree(hclust(congress, method = "average"), k = 2)
> names[g==1]
[1]   "Hunt(R)"              "Sandman(R)"          "Freylinghuysen(R)"
[4]   "Forsythe(R)"          "Widnall(R)"          "Maraziti(R)"
> names[g==2]
[1]   "Howard(D)"   "Thompson(D)"    "Roe(D)"        "Heltoski(D)"
[5]   "Rodino(D)"   "Minish(D)"      "Rinaldo(R)"    "Daniels(D)"
[9]   "Patten(D)"
```

The plot is shown in Figure 10.7. The Democrat/Republican division is clearly visible, as is the fact that Sandman is a distinct outlier. In fact, this particular congressperson had a greater tendency to abstain than the others.

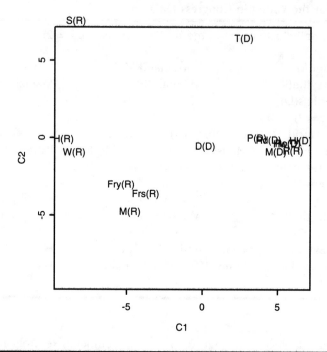

Figure 10.7 Two-dimensional solution from classical multidimensional scaling applied to voting behaviour data.

Exercises

10.1. A suggestion that has been made for deciding on the number of groups when using *k*-means clustering is to plot the values of within-group variance against the number of clusters and to take the number corresponding to any distinct 'elbow' in this plot. Investigate this possibility for the lowest temperature data.

10.2. Display the lowest temperature data in the space of their first two principal component scores, indicating on the plot the three-group cluster solution from *k*-means.

10.3. Investigate whether the two-dimensional solution for the voting data from classical multidimensional scaling is adequate for the data.

Chapter 11

Bivariate Density Estimation and Discriminant Analysis: Blood Fat Concentration

11.1 Description of Data

In a study of 371 males with chest pain, data were collected on the concentration of plasma cholesterol and plasma triglycerides (mg/dl). The data are shown in Table 11.1. For 51 patients there was no evidence of heart disease; for the remaining 320 there was evidence of narrowing of the arteries. Interest lies in assessing whether the two measurements made can be used to discriminate between men with and men without heart disease.

11.2 Bivariate Density Estimation and Discriminant Function Analysis

The data in Table 11.1 will first be explored by plotting a scattergram of the two variables and estimating their *bivariate density function*. Density estimation is described in detail in Silverman (1986), but for bivariate data it generally involves use of a *kernel estimator* of the form

Table 11.1 Blood Fat Data (Plasma cholesterol, plasma triglycerides)

51 patients with no evidence of heart disease

195,348	237,174	205,158	201,171	190,85	180,82	193,210
170,90	150,167	200,154	228,119	169,86	178,166	251,211
234,143	222,284	116,87	157,134	194,121	130,64	206,99
158,87	167,177	217,114	234,116	190,132	178,157	265,73
219,98	266,486	190,108	156,126	187,109	149,146	147,95
155,48	207,195	238,172	168,71	210,91	208,139	160,116
243,101	209,97	221,156	178,116	289,120	201,72	168,100
168,227	207,160					

320 patients with evidence of heart disease

184,145	263,142	185,115	271,128	173,56	230,304	222,151
215,168	233,340	212,171	221,140	239,97	168,131	231,145
221,432	131,137	211,124	232,258	313,256	240,221	176,166
210,92	251,189	175,148	185,256	184,222	198,149	198,333
208,112	284,245	231,181	171,165	258,210	164,76	230,492
197,87	216,112	230,90	265,156	197,158	230,146	233,142
250,118	243,50	175,489	200,68	240,196	185,116	213,130
180,80	208,220	386,162	236,152	230,162	188,220	200,101
212,130	193,188	230,158	169,112	181,104	189,84	180,202
297,232	232,328	150,426	239,154	178,100	242,144	323,196
168,208	197,291	417,198	172,140	240,441	191,115	217,327
208,262	220,75	191,115	119,84	171,170	179,126	208,149
180,102	254,153	191,136	176,217	283,424	253,222	220,172
268,154	248,312	245,120	171,108	239,92	196,141	247,137
219,454	159,125	200,152	233,127	232,131	189,135	237,400
319,418	171,78	194,183	244,108	236,148	260,144	254,170
250,161	196,130	298,143	306,408	175,153	251,117	256,271
285,930	184,255	228,142	171,120	229,242	195,137	214,223
221,268	204,150	276,199	165,121	211,91	264,259	245,446
227,146	197,265	196,103	193,170	211,122	185,120	157,59
224,124	209,82	223,80	278,152	251,152	140,164	197,101
172,106	174,117	192,101	221,179	283,199	178,109	185,168
181,119	191,233	185,130	206,133	210,217	226,72	219,267
215,325	228,130	245,257	186,273	242,85	201,297	239,137
179,126	218,123	279,317	234,135	264,269	237,88	162,91
245,166	191,90	207,316	248,142	139,173	246,87	247,91
193,290	332,250	194,116	195,363	243,112	271,89	197,347
242,179	175,246	138,91	244,177	206,201	191,149	223,154
172,207	190,120	144,125	194,125	105,36	201,92	193,259
262,88	211,304	178,84	331,134	235,144	267,199	227,202

Table 11.1 (Continued)   Blood Fat Data (Plasma cholesterol, plasma triglycerides)

243,126	261,174	185,100	171,90	222,229	231,161	258,328
211,306	249,256	209,89	177,133	165,151	299,93	274,323
219,163	233,101	220,153	348,154	194,400	230,137	250,160
173,300	260,127	258,151	131,61	168,91	208,77	287,209
308,260	227,172	168,126	178,101	164,80	151,73	165,155
249,146	258,145	194,196	140,99	187,390	171,135	221,156
294,135	167,80	208,201	208,148	185,231	159,82	222,108
266,164	217,227	249,200	218,207	245,322	242,180	262,169
169,158	204,84	184,182	206,148	198,124	242,248	189,176
260,98	199,153	207,150	206,107	210,95	229,296	232,583
267,192	228,149	187,115	304,149	140,102	209,376	198,105
270,110	188,148	160,125	218,96	257,402	259,240	139,54
213,261	178,125	172,146	198,103	222,348	238,156	273,146
131,96	233,141	269,84	170,284	149,237	194,272	142,111
218,567	194,278	252,233	184,184	203,170	239,38	232,161
225,240	280,218	185,110	163,156	216,101		

$$\hat{f}(x, y) = \frac{1}{nb_x b_y} \sum_{i=1}^{n} K\left[\frac{x - X_i}{b_x}, \frac{y - Y_i}{b_y}\right] \qquad (11.1)$$

where $(X_1, Y_1) \cdots (X_n, Y_n)$ are the sample bivariate observations, b_x and b_y are smoothing parameters usually known as *bandwidth*, and K is a *kernel function*, commonly, the standard bivariate normal density, i.e.,

$$K(x, y) = \frac{1}{2\pi} \exp\left[-\frac{1}{2}(x^2 + y^2)\right] \qquad (11.2)$$

In essence, the estimator is a sum of 'bumps' placed at the observations. The kernel function determines the shape of the bumps, and the bandwidth determines their width. A value often suggested for the bandwidth is $n^{-1/5}$ (see Venables and Ripley, 1999), for each dimension.

Following this initial exploration various discriminant function procedures will be applied to investigate the possibility of using the cholesterol and triglyceride measurements to discriminate between people with heart disease and those without. Details of such techniques are given in Everitt and Dunn (2001), but the simplest, linear discriminant function analysis consists of finding a linear function of the two variables that best distinguishes between members of the two groups. The coefficients that define this function, a_1 and a_2, can be shown to be given by

$$a = S^{-1}\left(\bar{x}_1 - \bar{x}_2\right) \qquad (11.3)$$

where $a' = [a_1, a_2]$, S is the estimator of the assumed common covariance matrix of the two groups, and $\bar{x}_1$ and $\bar{x}_2$ their mean vectors. Having found the vector, a, a discriminant score can be found for each individual as

$$z = a_1 \text{ Cholesterol} + a_2 \text{ Triglycerides} \qquad (11.4)$$

Individuals can now be assigned to one of the two groups on the basis of their discriminant score by comparing it to the threshold value $\frac{1}{2}(\bar{z}_1 + \bar{z}_2)$, where $\bar{z}_1$ and $\bar{z}_2$ are the mean discriminant scores of each group. If a score is larger than the threshold, the individual is classified into one group; if lower than the threshold, into the other. (This rule is based on the assumption that the *prior probabilities* of being in either group are the same. It is this rather unrealistic assumption that we shall make in this chapter, although in designing a classification rule for use in practice, realistic estimates of the prior probabilities would clearly be needed. For the adjusted classification rule in this case, see Everitt and Dunn, 2001.)

The performance of a discriminant analysis classification rule is assessed by estimating the *misclassification rate* of the rule. One possible estimator is the misclassification rate of the rule based on the data from which it was derived. This is, however, well known to be a very poor estimator and there are several alternatives that are preferred (see Hand, 1998, for details). The most commonly used of the available methods is the so-called leaving-one-out method, in which the discriminant function is derived from just $n - 1$ members of the sample and then used to classify the member left out. The process is repeated n times, leaving out each sample member in turn.

11.3 Analysis Using S-PLUS

The blood fat data are available as the data frame **bf**, the contents of which are shown in Table 11.2.

11.3.1 Bivariate Density Estimation

We begin by constructing a scatterplot of the data in which the members of the no heart disease group are identified by a minus sign and the members of the heart disease group by a plus sign.

Table 11.2   Contents of bf Data Frame

> *bf*

	Group	Cholesterol	Triglycerides
1	No heart disease	195	348
2	No heart disease	237	174
3	No heart disease	205	158
4	No heart disease	201	171
5	No heart disease	190	85
6	No heart disease	180	82
7	No heart disease	193	210
8	No heart disease	170	90
9	No heart disease	150	167
10	No heart disease	200	154
11	No heart disease	228	119
12	No heart disease	169	86
13	No heart disease	178	166
14	No heart disease	251	211
15	No heart disease	234	143
16	No heart disease	222	284
17	No heart disease	116	87
18	No heart disease	157	134
19	No heart disease	194	121
20	No heart disease	130	64
21	No heart disease	206	99
22	No heart disease	158	87
23	No heart disease	167	177
24	No heart disease	217	114
25	No heart disease	234	116
26	No heart disease	190	132
27	No heart disease	178	157
28	No heart disease	265	73
29	No heart disease	219	98
30	No heart disease	266	486
31	No heart disease	190	108
32	No heart disease	156	126
33	No heart disease	187	109
34	No heart disease	149	146
35	No heart disease	147	95
36	No heart disease	155	48
37	No heart disease	207	195
38	No heart disease	238	172
39	No heart disease	168	71

Table 11.2 (Continued)   Contents of bf Data Frame

> *bf*

	Group	Cholesterol	Triglycerides
40	No heart disease	210	91
41	No heart disease	208	139
42	No heart disease	160	116
43	No heart disease	243	101
44	No heart disease	209	97
45	No heart disease	221	156
46	No heart disease	178	116
47	No heart disease	289	120
48	No heart disease	201	72
49	No heart disease	168	100
50	No heart disease	168	227
51	No heart disease	207	160
52	Heart disease	184	145
53	Heart disease	263	142
54	Heart disease	185	115
55	Heart disease	271	128
56	Heart disease	173	56
57	Heart disease	230	304
58	Heart disease	222	151
59	Heart disease	215	168
60	Heart disease	233	340
61	Heart disease	212	171
62	Heart disease	221	140
63	Heart disease	239	97
64	Heart disease	168	131
65	Heart disease	231	145
66	Heart disease	221	432
67	Heart disease	131	137
68	Heart disease	211	124
69	Heart disease	232	258
70	Heart disease	313	256
71	Heart disease	240	221
72	Heart disease	176	166
73	Heart disease	210	92
74	Heart disease	251	189
75	Heart disease	175	148
76	Heart disease	185	256
77	Heart disease	184	222
78	Heart disease	198	149

Table 11.2 (Continued) Contents of **bf** Data Frame

> *bf*

	Group	Cholesterol	Triglycerides
79	Heart disease	198	333
80	Heart disease	208	112
81	Heart disease	284	245
82	Heart disease	231	181
83	Heart disease	171	165
84	Heart disease	258	210
85	Heart disease	164	76
86	Heart disease	230	492
87	Heart disease	197	87
88	Heart disease	216	112
89	Heart disease	230	90
90	Heart disease	265	156
91	Heart disease	197	158
92	Heart disease	230	146
93	Heart disease	233	142
94	Heart disease	250	118
95	Heart disease	243	50
96	Heart disease	175	489
97	Heart disease	200	68
98	Heart disease	240	196
99	Heart disease	185	116
100	Heart disease	213	130
101	Heart disease	180	80
102	Heart disease	208	220
103	Heart disease	386	162
104	Heart disease	236	152
105	Heart disease	230	162
106	Heart disease	188	220
107	Heart disease	200	101
108	Heart disease	212	130
109	Heart disease	193	188
110	Heart disease	230	158
111	Heart disease	169	112
112	Heart disease	181	104
113	Heart disease	189	84
114	Heart disease	180	202
115	Heart disease	297	232
116	Heart disease	232	328
117	Heart disease	150	426

Table 11.2 (Continued)   Contents of bf Data Frame

> *bf*

	Group	Cholesterol	Triglycerides
118	Heart disease	239	154
119	Heart disease	178	100
120	Heart disease	242	144
121	Heart disease	323	196
122	Heart disease	168	208
123	Heart disease	197	291
124	Heart disease	417	198
125	Heart disease	172	140
126	Heart disease	240	441
127	Heart disease	191	115
128	Heart disease	217	327
129	Heart disease	208	262
130	Heart disease	220	75
131	Heart disease	191	115
132	Heart disease	119	84
133	Heart disease	171	170
134	Heart disease	179	126
135	Heart disease	208	149
136	Heart disease	180	102
137	Heart disease	254	153
138	Heart disease	191	136
139	Heart disease	176	217
140	Heart disease	283	424
141	Heart disease	253	222
142	Heart disease	220	172
143	Heart disease	268	154
144	Heart disease	248	312
145	Heart disease	245	120
146	Heart disease	171	108
147	Heart disease	239	92
148	Heart disease	196	141
149	Heart disease	247	137
150	Heart disease	219	454
151	Heart disease	159	125
152	Heart disease	200	152
153	Heart disease	233	127
154	Heart disease	232	131
155	Heart disease	189	135
156	Heart disease	237	400

Table 11.2 (Continued)   Contents of bf Data Frame

> *bf*

	Group	Cholesterol	Triglycerides
157	Heart disease	319	418
158	Heart disease	171	78
159	Heart disease	194	183
160	Heart disease	244	108
161	Heart disease	236	148
162	Heart disease	260	144
163	Heart disease	254	170
164	Heart disease	250	161
165	Heart disease	196	130
166	Heart disease	298	143
167	Heart disease	306	408
168	Heart disease	175	153
169	Heart disease	251	117
170	Heart disease	256	271
171	Heart disease	285	930
172	Heart disease	184	255
173	Heart disease	228	142
174	Heart disease	171	120
175	Heart disease	229	242
176	Heart disease	195	137
177	Heart disease	214	223
178	Heart disease	221	268
179	Heart disease	204	150
180	Heart disease	276	199
181	Heart disease	165	121
182	Heart disease	211	91
183	Heart disease	264	259
184	Heart disease	245	446
185	Heart disease	227	146
186	Heart disease	197	265
187	Heart disease	196	103
188	Heart disease	193	170
189	Heart disease	211	122
190	Heart disease	185	120
191	Heart disease	157	59
192	Heart disease	224	124
193	Heart disease	209	82
194	Heart disease	223	80
195	Heart disease	278	152

Table 11.2 (Continued)   Contents of bf Data Frame

> bf

	Group	Cholesterol	Triglycerides
196	Heart disease	251	152
197	Heart disease	140	164
198	Heart disease	197	101
199	Heart disease	172	106
200	Heart disease	174	117
201	Heart disease	192	101
202	Heart disease	221	179
203	Heart disease	283	199
204	Heart disease	178	109
205	Heart disease	185	168
206	Heart disease	181	119
207	Heart disease	191	233
208	Heart disease	185	130
209	Heart disease	206	133
210	Heart disease	210	217
211	Heart disease	226	72
212	Heart disease	219	267
213	Heart disease	215	325
214	Heart disease	228	130
215	Heart disease	245	257
216	Heart disease	186	273
217	Heart disease	242	85
218	Heart disease	201	297
219	Heart disease	239	137
220	Heart disease	179	126
221	Heart disease	218	123
222	Heart disease	279	317
223	Heart disease	234	135
224	Heart disease	264	269
225	Heart disease	237	88
226	Heart disease	162	91
227	Heart disease	245	166
228	Heart disease	191	90
229	Heart disease	207	316
230	Heart disease	248	142
231	Heart disease	139	173
232	Heart disease	246	87
233	Heart disease	247	91
234	Heart disease	193	290

Table 11.2 (Continued)   Contents of bf Data Frame

> *bf*

	Group	Cholesterol	Triglycerides
235	Heart disease	332	250
236	Heart disease	194	116
237	Heart disease	195	363
238	Heart disease	243	112
239	Heart disease	271	89
240	Heart disease	197	347
241	Heart disease	242	179
242	Heart disease	175	246
243	Heart disease	138	91
244	Heart disease	244	177
245	Heart disease	206	201
246	Heart disease	191	149
247	Heart disease	223	154
248	Heart disease	172	207
249	Heart disease	190	120
250	Heart disease	144	125
251	Heart disease	194	125
252	Heart disease	105	36
253	Heart disease	201	92
254	Heart disease	193	259
255	Heart disease	262	88
256	Heart disease	211	304
257	Heart disease	178	84
258	Heart disease	331	134
259	Heart disease	235	144
260	Heart disease	267	199
261	Heart disease	227	202
262	Heart disease	243	126
263	Heart disease	261	174
264	Heart disease	185	100
265	Heart disease	171	90
266	Heart disease	222	229
267	Heart disease	231	161
268	Heart disease	258	328
269	Heart disease	211	306
270	Heart disease	249	256
271	Heart disease	209	89
272	Heart disease	177	133
273	Heart disease	165	151

Table 11.2 (Continued)   Contents of bf Data Frame

> bf

	Group	Cholesterol	Triglycerides
274	Heart disease	299	93
275	Heart disease	274	323
276	Heart disease	219	163
277	Heart disease	233	101
278	Heart disease	220	153
279	Heart disease	348	154
280	Heart disease	194	400
281	Heart disease	230	137
282	Heart disease	250	160
283	Heart disease	173	300
284	Heart disease	260	127
285	Heart disease	258	151
286	Heart disease	131	61
287	Heart disease	168	91
288	Heart disease	208	77
289	Heart disease	287	209
290	Heart disease	308	260
291	Heart disease	227	172
292	Heart disease	168	126
293	Heart disease	178	101
294	Heart disease	164	80
295	Heart disease	151	73
296	Heart disease	165	155
297	Heart disease	249	146
298	Heart disease	258	145
299	Heart disease	194	196
300	Heart disease	140	99
301	Heart disease	187	390
302	Heart disease	171	135
303	Heart disease	221	156
304	Heart disease	294	135
305	Heart disease	167	80
306	Heart disease	208	201
307	Heart disease	208	148
308	Heart disease	185	231
309	Heart disease	159	82
310	Heart disease	222	108
311	Heart disease	266	164
312	Heart disease	217	227

Table 11.2 (Continued) Contents of bf Data Frame

> *bf*

	Group	Cholesterol	Triglycerides
313	Heart disease	249	200
314	Heart disease	218	207
315	Heart disease	245	322
316	Heart disease	242	180
317	Heart disease	262	169
318	Heart disease	169	158
319	Heart disease	204	84
320	Heart disease	184	182
321	Heart disease	206	148
322	Heart disease	198	124
323	Heart disease	242	248
324	Heart disease	189	176
325	Heart disease	260	98
326	Heart disease	199	153
327	Heart disease	207	150
328	Heart disease	206	107
329	Heart disease	210	95
330	Heart disease	229	296
331	Heart disease	232	583
332	Heart disease	267	192
333	Heart disease	228	149
334	Heart disease	187	115
335	Heart disease	304	149
336	Heart disease	140	102
337	Heart disease	209	376
338	Heart disease	198	105
339	Heart disease	270	110
340	Heart disease	188	148
341	Heart disease	160	125
342	Heart disease	218	96
343	Heart disease	257	402
344	Heart disease	259	240
345	Heart disease	139	54
346	Heart disease	213	261
347	Heart disease	178	125
348	Heart disease	172	146
349	Heart disease	198	103
350	Heart disease	222	348
351	Heart disease	238	156

Table 11.2 (Continued)   Contents of bf Data Frame

> bf

	Group	Cholesterol	Triglycerides
352	Heart disease	273	146
353	Heart disease	131	96
354	Heart disease	233	141
355	Heart disease	269	84
356	Heart disease	170	284
357	Heart disease	149	237
358	Heart disease	194	272
359	Heart disease	142	111
360	Heart disease	218	567
361	Heart disease	194	278
362	Heart disease	252	233
363	Heart disease	184	184
364	Heart disease	203	170
365	Heart disease	239	38
366	Heart disease	232	161
367	Heart disease	225	240
368	Heart disease	280	218
369	Heart disease	185	110
370	Heart disease	163	156
371	Heart disease	216	101

```
attach(bf)
plot(Cholesterol,Triglycerides,type="n")
text(Cholesterol,Triglycerides,labels=
rep("-","+"),c(51,320)))
legend(locator(1),c("No heart disease","Heart
disease"),pch="-+")
```

The resulting diagram is shown in Figure 11.1. The main features in this diagram are the considerable overlap between the two groups and the presence of an individual with an extremely high triglyceride value. This is individual number 171. Since the value is so extreme, this individual will be removed from further analyses.

A first, very crude, estimate of the bivariate density of the two variables can be obtained using the **hist2d** function in S-PLUS, which simply counts the number of observations in each cell of a grid constructed for the data. The resulting counts can be viewed graphically using the **persp** function.

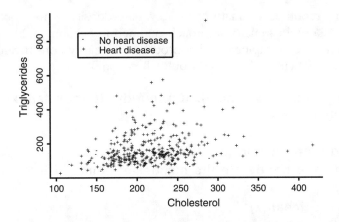

Figure 11.1   Scatterplot of cholesterol and triglycerides identifying those men with heart disease and those without.

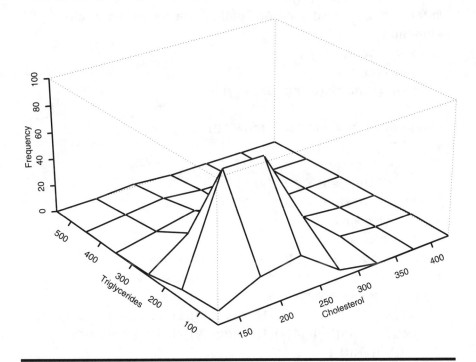

Figure 11.2   Two-dimensional histogram of blood fat data.

```
persp(hist2d(Cholesterol[-171],Triglycerides [-171]),
zlab="Frequency",xlab="Cholesterol",ylab="Triglycerides")
```

The resulting diagram is shown in Figure 11.2.

A far 'smoother' estimate of the bivariate density can be obtained by using the kernel estimator approach outlined in Section 11.2. A relatively crude function to implement this procedure is easily constructed making use of, in particular, the S-PLUS **outer** function.

```
kden<-function(x, y, ngridx=30, ngridy=30, constant.x
=1, constant.y=1)
    {
#x and y are vectors containing the bivariate data
#ngridx and ngridy are the number of points in the grid
#
    mx <- mean(x)
    sdx <- sqrt(var(x))
    my <- mean(y)
    sdy <- sqrt(var(y))
#standardize x and y before estimation using the scale
#function
    x <- scale(x)
    y <- scale(y)
    den <- matrix(0, ngridx, ngridy)
#
#find possible value for bandwidth
#
    n <- length(x)
    hx <- constant.x * n^(-0.2)
    hy <- constant.y * n^(-0.2)
    h <- hx * hy
    hsqrt <- sqrt(h)
#set up grid at which to calculate densities
#use the outer function to compute densities
#and accumulate over the n observations
    seqx <- seq(range(x)[1], range(x)[2], length=ngridx)
    seqy <- seq(range(y)[1], range(y)[2], length=ngridy)
    for(i in 1:n) {
        X <- x[i]
        Y <- y[i]
        xx <- (seqx - X)/hsqrt
        yy <- (seqy - Y)/hsqrt
        den <- den + outer(xx, yy, function(x, y)
```

```
                    exp(-0.5 * (x^2 + y^2)))
                }
        den <- den/(n * 2 * pi * h)
        seqx <- sdx * seqx + mx
        seqy <- sdy * seqy + my
        result <- list(seqx=seqx, seqy=seqy, den=den)
        result
}
```

We can now use the function **kden** to estimate the required bivariate density and then view the estimate using **persp.**

```
den<-kden(Cholesterol[-171],Triglycerides[-171])
persp(den$seqx,den$seqy,den$den,xlab="Cholesterol",ylab=
"Triglycerides",zlab="Density")
```

The resulting diagram appears in Figure 11.3. The bivariate density looks relatively 'normal' apart from perhaps a slight 'bumpiness' for triglyceride values above 200, but there is no obvious bimodality that would indicate relatively clear separation between the with and without heart disease groups.

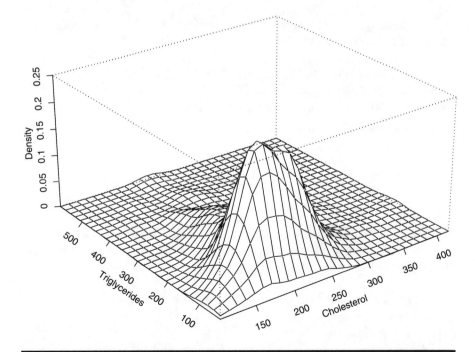

Figure 11.3   Kernel estimate of bivariate density function of blood fat data.

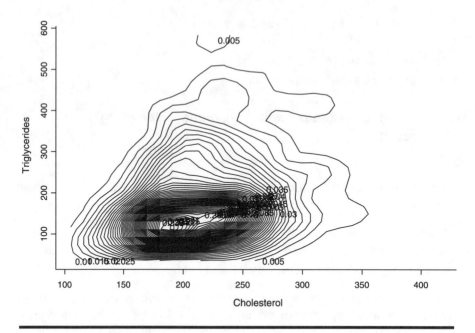

Figure 11.4   Contour plot of estimated bivariate density of blood fat data.

The density estimate might also be displayed as a contour plot using the S-PLUS **contour** function:

```
contour(den$seqx,den$seqy,den$den,xlab="Cholesterol",
ylab="Triglycerides",nlevels=50)
#the nlevels argument is used to increase
#the number of contours plotted
```

The contour plot is shown in Figure 11.4.

11.3.2 Discriminant Analysis

We begin this subsection by constructing the linear discriminant function for the no heart disease/heart disease groups first from principles using, in particular, the **solve** function in S-PLUS to invert a matrix:

```
#Calculate the discriminant function coefficients
#from first principles
#
#first calculate the covariance matrices in each group
```

```
#
#
bf <- bfc[-171, 2:3]
Group <- Group[-171]
v1 <- var(bf[Group=="No heart disease",])
v2 <- var(bf[Group=="Heart disease",])
#
#Calculate group mean vectors
#
m1 <- apply(bf[Group=="No heart disease",], 2, mean)
m2 <- apply(bf[Group=="Heart disease",], 2, mean)
#
#Find number of observations in each group
#
n1 <- length(Group[Group=="No heart disease"])
n2 <- length(Group[Group=="Heart disease"])
#
#Estimate assumed common covariance matrix
#
v <- ((n1 - 1) * v1 + (n2 - 1) * v2)/(n1 + n2 - 2)
#
#Calculate inverse of common covariance matrix and store
#as v
#
v <- solve(v)
#
#calculate discriminant function coefficients
#
a <- v %*% (m1 - m2)
#
#calculate threshold value for discriminant rule
#
z12 <- (m1 %*% a + m2 %*% a)/2
#
#plot the data labelling the members of the two groups
#and add the discriminant function line and an
#appropriate legend
#
```

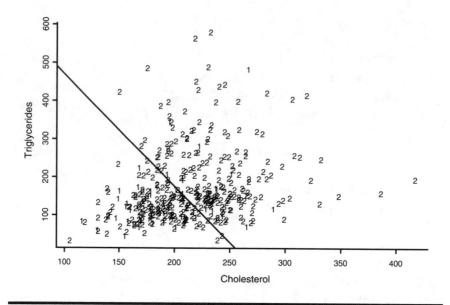

Figure 11.5 Plotted and discriminant function. 1 = No heart disease; 2 = heart disease.

```
plot(bf[, 1], bf[, 2], xlab="Cholesterol", ylab=
"Tryglicerides", type="n")
text(bf[, 1], bf[, 2], labels=Group)
legend(locator(1), c("No heart disease", "Heart
disease"), pch="12")
#
#add estimated discriminant function to scatter plot
#
abline(z12/a[2], - a[1]/a[2], lwd=2, lty=1)
```

The estimated discriminant function is

$$z = -0.01007 \text{ Cholesterol} -0.00343 \text{ Triglycerides} \qquad (11.5)$$

and the threshold value is -2.62.

The plotted data and discriminant function are shown in Figure 11.5. Observations falling above the line would be classified as 'heart disease', those below the line as 'no heart disease'.

We can also obtain the discriminant function along with information about its performance in terms of estimated misclassification rate by using the **discrim** function in S-PLUS.

```
bf.disc<-discrim(Group~Cholesterol+Triglycerides,data
=bf,prior="uniform")
summary(bf.disc)
```

The results are given in Table 11.3. The discriminant function coefficients given in (11.5) are obtained from this table by simply subtracting the appropriate linear coefficients for no heart disease and heart disease. The test for homogeneity of variance is not quite significant at the 5% level, and the tests for a difference in the bivariate means of the two groups all take the same p-value since there are only two groups. The difference is highly significant. Here, the estimated misclassification rates estimated from the original sample and from the leaving-out-one method are very similar.

The linear discriminant function derived above assumes that both groups have the same population covariance matrix. When this assumption is thought to be doubtful, a quadratic discriminant function can be derived (see Everitt and Dunn, 2001, for details). Largely as an exercise, we shall demonstrate how to find this function for the blood fat data by using the S-PLUS **Discriminant Analysis** dialog. For this analysis we shall use all the data, i.e., now including individual 171.

- ■ Click on **Statistics**.
- ■ Select **Multivariate**.
- ■ Select **Discriminant Analysis**.

In the **Discriminant Analysis** dialog, which now appears, choose **bf** as the data set, and then Group as the **Dependent variable** and Cholesterol and Triglycerides as the **Independent**. Select **heteroscedastic** for **Covariance Struct** and set **Group Prior** to uniform. The dialog now appears as shown in Figure 11.6. Now check the **Results** tab and tick **Plot**. Finally, click on **OK**.

The results of the analysis are shown in Table 11.4 and the plotted quadratic discriminant function is shown in Figure 11.7. In Table 11.4, there are quadratic coefficients in addition to linear coefficients. The test for homogeneity of covariances differs from that in Table 11.3 because of the inclusion of individual 171 in this analysis. With this individual included, there is a significant difference between the covariance matrices of the two groups. The quadratic discriminant function classifies more of the no heart disease correctly than the linear function, but does less well on the heart disease group.

Table 11.3   Results from Using discrim Function on Blood Fat Data

Call:
discrim(Group ~ Cholesterol + Triglycerides, data = bf, prior =
    "uniform")

Group means:

	Cholesterol	Triglycerides	N	Priors
No heart disease	195	140	51	0.5
Heart disease	216	177	31	0.5

Covariance Structure: homoscedastic

	Cholesterol	Triglycerides
Cholesterol	1768	844
Triglycerides	8205	

Constants:

No heart disease	Heart disease
–11.6	–14.2

Linear Coefficients:

	No heart disease	Heart disease
Cholesterol	0.108	0.118
Triglycerides	0.006	0.009

Tests for Homogeneity of Covariances:

	Statistic	df	Pr
Box.M	7.73	3	0.052
adj.M	7.62	3	0.055

Tests for the Equality of Means:
Group Variable: Group

	Statistics	F	df1	df2	Pr
Wilks Lambda	0.9616	7.33	2	367	0.000758
Pillai Trace	0.0384	7.33	2	367	0.000758
Hoteling-Lawley Trace	0.0399	7.33	2	367	0.000758
Roy Greatest Root	0.0399	7.33	2	367	0.000758

* Tests assume covariance homoscedasticity.
   F Statistic for Wilks' Lambda is exact.
   F Statistic for Roy's Greatest Root is an upper bound.

Table 11.3 (Continued)   Results from Using discrim Function on Blood Fat Data

Hotelling's T Squared for Differences in Means Between Each Group:

	F	df1	df2	Pr
No heart disease-Heart disease	7.33	2	367	0.000758

95% Simultaneous Confidence Intervals Using the Dunnett Method:

	Estimate	Std.Error	Lower Bound
Nhrtdiss.Cholestr-Hrtdises.Cholestr	–20.7	6.34	–34.9
Nhrtdiss.Trglycrd-Hrtdises.Trglycrd	–36.6	13.70	–67.3

	Upper Bound	
Nhrtdiss.Cholestr-Hrtdises.Cholestr	–6.49	****
Nhrtdiss.Trglycrd-Hrtdises.Trglycrd	–6.03	****

(critical point: 2.2411)
* Intervals excluding 0 are flagged by '****'

Mahalanobis Distance:

	No heart disease	Heart disease
No heart disease	0.000	0.334
Heart disease		0.000

Kolmogorov-Smirnov Test for Normality:

	Statistic	Probability
Cholesterol	0.057	0.180
Triglycerides	0.048	0.362

Plug-in classification table:

	No heart disease	Heart disease	Error	Posterior.Error
No heart disease	35	16	0.314	0.461
Heart disease	134	185	0.420	0.322
Overall			0.367	0.391

	Stratified.Error
No heart disease	0.340
Heart disease	0.451
Overall	0.396

(from = rows,to = columns)

Optimal Error Rate:

No heart disease	Heart disease	overall
0.449	0.449	0.449

Table 11.3 (Continued)   Results from Using discrim Function on Blood Fat Data

Rule Mean Square Error: 0.462
(conditioned on the training data)

Cross-validation table:

	No heart disease	Heart disease	Error	Posterior.Error
No heart disease	34	17	0.333	0.458
Heart disease	136	183	0.426	0.324
Overall			0.380	0.391

	Stratified.Error
No heart disease	0.349
Heart disease	0.441
Overall	0.395

(from = rows,to = columns)

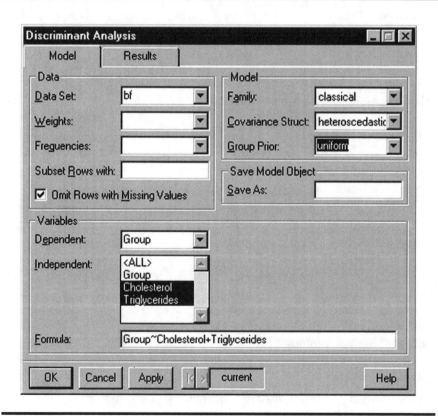

Figure 11.6   Completed Discriminant Analysis dialog.

Table 11.4   Quadratic Discriminant Function Results for Blood Fat Data

*** Discriminant Analysis ***

Call:
discrim(Group ~ Cholesterol + Triglycerides, data = bf, family =
    Classical(cov.structure = "heteroscedastic"), na.action =
    na.omit, prior = "uniform")

Group means:

	Cholesterol	Triglycerides	N	Priors
No heart disease	195	140	51	0.5
Heart disease	216	179	320	0.5

Covariance Structure: heteroscedastic

Group: No heart disease

	Cholesterol	Triglycerides
Cholesterol	1304	873
Triglycerides		5519

Group: Heart disease

	Cholesterol	Triglycerides
Cholesterol	1850	999
Triglycerides		10373

Constants:

No heart disease	Heart disease
–23.2	–21.9

Linear Coefficients:

	No heart disease	Heart disease
Cholesterol	0.148	0.113
Triglycerides	0.002	0.006

Quadratic coefficents:

group: No heart disease

	Cholesterol	Triglycerides
Cholesterol	–0.000429	0.000068
Triglycerides		–0.000101

Table 11.4 (Continued)   Quadratic Discriminant Function Results for Blood Fat Data

group: Heart disease

	Cholesterol	Triglycerides
Cholesterol	–0.000285	0.0000274
Triglycerides		–0.0000508

Tests for Homogeneity of Covariances:

	Statistic	df	Pr
Box.M	11.2	3	0.011
adj.M	11.1	3	0.011

Hotelling's T Squared for Differences in Means between Each Group:

	F	df1	df2	Pr
No heart disease-Heart disease	9.4	2	74.8	0.000228

* df2 is Yao's approximation.

95% Simultaneous Confidence Intervals Using the Dunnett Method:

	Estimate	Std.Error	Lower Bound
Nhrtdiss.Cholestr-Hrtdises.Cholestr	–20.9	5.6	–33.6
Nhrtdiss.Trglycrd-Hrtdises.Trglycrd	–39.0	11.9	–65.9

	Upper Bound	
Nhrtdiss.Cholestr-Hrtdises.Cholestr	–8.2	****
Nhrtdiss.Trglycrd-Hrtdises.Trglycrd	–12.1	****

(critical point: 2.272)
* Intervals excluding 0 are flagged by '****'

Pairwise Generalized Squared Distances:

	No heart disease	Heart disease
No heart disease	0.000	0.315
Heart disease	0.462	0.000

Kolmogorov-Smirnov Test for Normality:

	Statistic	Probability
Cholesterol	0.062	0.111
Triglycerides	0.043	0.505

Table 11.4 (Continued)   Quadratic Discriminant Function Results for Blood Fat Data

Plug-in classification table:

	No heart disease	Heart disease	Error	Posterior.Error
No heart disease	42	9	0.176	0.246
Heart disease	177	143	0.553	0.407
Overall		0.365	0.326	

	Stratified.Error
No heart disease	0.111
Heart disease	0.552
Overall	0.331

(from = rows, to = columns)

Rule Mean Square Error: 0.522
(conditioned on the training data)

Cross-validation table:

	No heart disease	Heart disease	Error	Posterior.Error
No heart disease	40	11	0.216	0.252
Heart disease	177	143	0.553	0.401
Overall		0.384	0.326	

	Stratified.Error
No heart disease	0.137
Heart disease	0.526
Overall	0.331

(from = rows, to = columns)

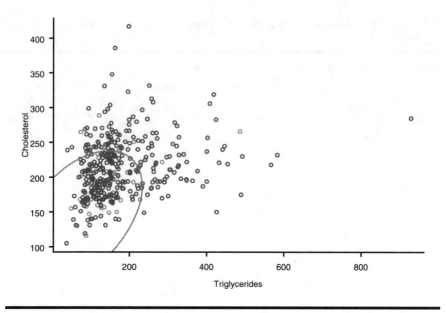

Figure 11.7   Plotted quadratic discriminant function.

Exercises

11.1.   Investigate what happens when the **discrim** function is used on the blood fat data with other than uniform priors.

11.2.   Investigate using the **discrim** function on the blood fat data when the assumption about the group covariance matrix is neither that they are equal nor that they are completely different.

Appendix A

The S-PLUS Language

A1   Vectors and Matrices

Vectors can be created using the concatenate function, **c**, which combines all values given as arguments to the function into a vector.

```
>x<-c(1,2,3,4)
>y<-c(5,6,7,8)
>xy<-c(x,y)
>xy
[1] 1 2 3 4 5 6 7 8
```

The number of elements in a vector can be determined using the **length** function

```
>length(xy)
[1] 8
```

Particular elements of vectors can be accessed using the square bracket nomenclature

```
>xy[2]
[1] 2
>xy[c(1,3,8)]
[1] 1  3  8
>xy[-c(1,3,8)]
[1] 2 4 5 6 7
```

The **c** function can also be used to combine strings denoted by enclosing them in " ". For example:

```
>names<-c("Brian","Rachel","Hywel","Dafydd")
>names
[1] "Brian", "Rachel", "Hywel", "Dafydd"
```

The **c** function also works with a mixture of numeric and string values, but in this case all elements in the resulting vector will be converted to strings:

```
>xnames<-c(x,names)
>xnames
[1] "1" "2" "3" "4" "Brian" "Rachel" "Hywel"
[8] "Dafydd"
```

Arithmetic operators (see Table A1) work element-wise on vectors; so, for example,

```
>x*y
[1] 5 12 21 32
>x/x
[1] 1 1 1 1
>x-y
[1]-4 -4   -4   -4
>2*x
[1]2 4 6 8
```

Table A1   Arithmetic Operators

Operator	Meaning	Expression	Result
+	plus	2 + 3	5
−	minus	5 − 2	3
*	times	5 * 2	10
/	divided by	10/2	5
^	power	2^3	8

We can also apply mathematical functions to vectors, and they will be applied element by element. (The most common of these functions are listed in Table A2.) To restrict the number of decimal points printed, we first use the **options** function as follows:

Table A2   Common Functions

S-PLUS Function	Meaning
sqrt	square root
log	natural logarithm
log10	logarithm base 10
exp	exponential
abs	absolute value
round	round to nearest integer
ceiling	round up
floor	round down
sin, cos, tan	sine, cosine, tangent
asin, acos, atan	arc sine, arc cosine, arc tangent

```
options(digits=3)
>log(xy)
[1] 0.000 0.693 1.099 1.386 1.609 1.792 1.946 2.079
>exp(xy)
[1] 2.72 7.39 20.09 54.60 148.41 403.43 1096.63 2980.96
>sqrt(x*x+y*y)
[1] 5.10 6.32 7.62 8.94
```

Matrix objects are also frequently required when using S-PLUS and can be created by use of the **matrix** function. The general syntax is

```
matrix(data, nrow, ncol, byrow=F)
```

The last argument specifies whether the matrix is to be filled row by row or column by column, with the latter being the default. Some examples will help to clarify how the **matrix** function operates:

```
>X<-matrix(c(x,y),nrow=2)
>X
      [,1]  [,2]  [,3]  [,4]
[1,]    1     3     5     7
[2,]    2     4     6     8
```

Here, the number of columns is not specified and so is determined by simple division.

```
>X<-matrix(c(x,y),nrow=2,byrow=T)
>X
      [,1]  [,2]  [,3]  [,4]
[1,]    1     2     3     4
[2,]    5     6     7     8
```

Here, the matrix is filled row-wise instead of by columns since **byrow** is set to true.

Again, the square bracket convention can be used to refer to particular elements of matrices, or particular rows or columns, or particular subsets of rows and columns.

```
>X[1,1]
[1]   1
>X[2,]
[1]  5   6   7   8
>X[,3]
[1]  3   7
>X[,c(1, 4)]
      [,1]   [,2]
[1,]    1      4
[2,]    5      8
```

As with vectors, arithmetic operations operate element by element when applied to matrices:

```
>X+X
      [,1]  [,2]  [,3]  [,4]
[1,]    2     4     6     8
[2,]   10    12    14    16
```

```
>X/X
      [,1]  [,2]  [,3]  [,4]
[1,]    1     1     1     1
[2,]    1     1     1     1
```

```
>sqrt(X)
      [,1]   [,2]   [,3]   [,4]
[1,]  1.00   1.41   1.73   2.00
[2,]  2.24   2.45   2.65   2.83
```

Matrix multiplication is performed using the %*% operator

```
>X%*%t(X)
        [,1]    [,2]
[1,]     30      70
[2,]     70     174
```

Here, the matrix **X** is multiplied by its transpose which is found using the t function. Trying to multiply **X** by itself would lead to an error message:

```
>X%*%X
Error in "%*%.default" (X,X): Number of columns of x
should be the same as number of rows of y
```

A very powerful procedure in S-PLUS is the ability to subset matrices and vectors using logical expressions (the symbols used for logical operations are listed in Table A3). In addition to logical operations, there are also a number of logical functions, i.e., functions that return the values T or F.

```
>is.numeric(4)
[1]   T
>is.character(4)
[1]   F
>is.character("B")
[1]   T
```

Table A3   Logical Operators

Operator	Meaning
<	less than
>	greater than
< =	less than or equal to
> =	greater than or equal to
==	equal to
! =	not equal to
&	and
		or
!	not

Of particular importance here is the function used for testing for missing values (NA in S-PLUS), is.na:

```
>is.na(c(10,20,NA,NA,3,NA))
[1] F F T T F T
>!is.na(c(10,20,NA,NA,3,NA))
[1] T T F F T F
```

Logical operators or functions operate on vectors and matrices in the same way as other operators, i.e., element by element:

```
>x<-c(1,2,3,4,5,6)
>x<4
[1] T T T F F F
```

A logical vector can be used to extract a subset of elements from another vector, for example,

```
>x[x<4]
[1] 1  2  3
>y<-c(2,8,10,12,16,18)
>y[x>=4]
[1] 12  16  18
>X<-matrix(c(1,2,-9,-9,7,-9,8,6),ncol=2)
>X[X==-9]<-NA
>X
      [,1]   [,2]
[1,]    1     7
[2,]    2    NA
[3,]   NA     8
[4,]   NA     6
```

A2 List Objects

List objects allow other S-PLUS objects to be linked together. For example,

```
>x<-c(1,2,3,4)
>X<-matrix(c(x,10,20,30,40),ncol=2)
>xXlist<-list(x,X)
>xXlist
```

```
[[1]]:
[1]  1   2   3   4
[[2]]:
        [,1]   [,2]
[1,]     1     10
[2,]     2     20
[3,]     3     30
[4,]     4     40
```

Note that the elements of the list are referred to by a double square bracket notation. So,

```
>xXlist[[2]]
        [,1]   [,2]
[1,]     1     10
[2,]     2     20
[3,]     3     30
[4,]     4     40
```

The components of a list can also be given names and can, of course, also include other list objects.

```
>newlist<-list(first=x,second=X,third=xXlist)
```

Now, the components of the list can also be referred to using the **list$name** notation

```
>newlist$first
[1]  1   2   3   4
```

```
>newlist$third
[[1]]:
[1]  1   2   3   4

[[2]]:
        [,1]   [,2]
[1,]     1     10
[2,]     2     20
[3,]     3     30
[4,]     4     40
```

A list object can be used to give labels to the rows and/or columns of a matrix which is often very useful. For example:

```
>amatrix<-matrix(c(1,2,3,10,20,30,0,0,0),nrow=3,byrow=T)
>amatrix
      [,1]  [,2]  [,3]
[1,]    1     2     3
[2,]   10    20    30
[3,]    0     0     0

>dimnames(amatrix)<-list(c("R1","R2","R3"),c("C1","C2","C3"))
>amatrix
      C1   C2   C3
R1     1    2    3
R2    10   20   30
R3     0    0    0
```

If the matrix was large, say, 500 rows and 50 columns, then helpful row and column labels can be created easily using the **paste** function:

```
>dimnames(amatrix)<-list(paste("R", 1:500), paste("C", 1:50))
```

The row labels will be R1, R2, ... R500, and the column labels C1, C2, ... C50.

A3   Data Frames

Data sets in S-PLUS are usually stored as data frames, since these can bind together vectors of different types (for example, numeric and character), retaining the correct type for each vector. In many respects a data frame is similar to a matrix so that each vector should have the same number of elements. The syntax for creating data frame objects is **data.frame(vector 1,vector 2,...)**; here is a small example:

```
>height<-c(50, 70, 45, 80, 100)
>weight<-c(120, 140, 100, 200, 190)
>age<-c(20, 40, 41, 31, 33)
>names<-c("Bob", "Ted", "Alice", "Mary", "Sue")
>sex<-c("Male", "Male", "Female", "Female", "Female")
>data<-data.frame (names, sex, height, weight, age)
```

```
>data
    names   sex       height  weight  age
1   Bob     Male      50      120     20
2   Ted     Male      70      140     40
3   Alice   Female    45      100     41
4   Mary    Female    80      200     31
5   Sue     Female    100     190     33
```

Particular parts of a data frame can be extracted in the same way as for matrices.

```
>data[,c(1, 2, 5)]
    names   sex       age
1   Bob     Male      20
2   Ted     Male      40
3   Alice   Female    41
4   Mary    Female    31
5   Sue     Female    33
```

Column names can also be used:
```
>data[,c("names","age")]
    names   age
1   Bob     20
2   Ted     40
3   Alice   41
4   Mary    31
5   Sue     33
```

Variables can also be accessed as with list objects:

```
>data$age
[1]  20  40  41  31  33
```

It is, however, more convenient to 'attach' a data frame object and then work with the column names directly. For example,

```
>attach(data)
>age
[1]  20  40  41  31  33
```

Note that the **attach** command places the data frame in the second position in the search path, so if, for example, we now assign a value to age

```
>age<-10
>age
[1]  10
```

a new object is created in the first position of the search path that 'masks' the age variable of the **data** data frame. Variables can be removed from the first position in the search path using the **rm** function.

```
>rm(age)
>age
1    2    3    4    5
20   40   41   31   33
```

To change the value of any or all elements of **age** in the **data** data frame, use the syntax

```
>data$age<-c(20,30,45,32,32)
```

A4 Reading Data into S-PLUS

S-PLUS can read data from a large number of different statistical packages and spreadsheet/database packages, including SPSS, Stata, SAS, Microsoft Excel, and Microsoft Access. To read a file from another package using the GUI, for example,

- Click on **File**.
- Select **Import Data**.
- Select **From File**.

The **Import Data** dialog as shown in Figure A1 now appears.

In the **File name** window, enter the file name and then select the appropriate type in the Files type window, e.g., SPSS files (*sav).

ASCII data can be read using the **scan** function. For example, if the file **data.dat** in the directory **C:\users\me** contains the matrix data

```
123
456
321
```

they can be read into a vector using **scan**, and then converted into a matrix using the **matrix** function as usual.

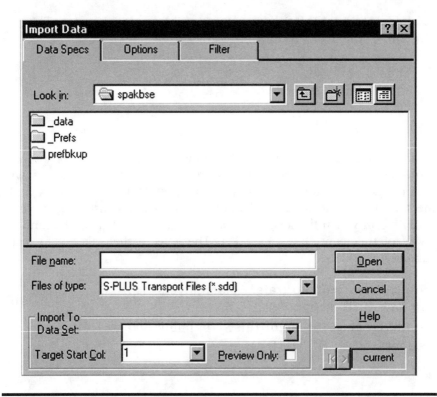

Figure A1   Import Data dialog.

```
>A<-matrix(scan(file="c:\\users\\me\\data.dat"),nrow=3,
byrow=T)
```

Note that the scan function reads the data in lexicographical order (down the rows), so that the **byrow=T** option needs to be used in the matrix function.

Perhaps the most convenient way of reading-in ASCII data is by using the **read.table** function which is used in creating data frames. If the first row of the file contains the variable names, the **header=T** option causes these names to be used for the columns of the data frame. If, for example, the data in **C:\users\me\children.dat** contain

```
age   sex   y
11    boy   3.2
9     boy   5.8
13    girl  1.2
```

they can be read-in using

```
>child<-read.table("c:\\users\\me\\children.dat",header=T)
>child
```

```
   age   sex   y
1  11    boy   3.2
2  9     boy   5.8
3  13    girl  1.2
```

In many cases, all variables will be initially read into data in numeric form and then categorical variables designated as factor variables with their categories suitably labelled by use of the **factor** function. For example, the **rats** data frame used in Chapter 3 was initially constructed from an ASCII file in which the levels of poison were coded 1, 2, and 3, and the levels of treatment, 1, 2, 3, and 4. These were then altered to factor variables as follows;

```
>rats$Poison<-factor(rats$Poison,levels=1:3,
>labels=c("P1","P2","P3"))
>rats$Treatment<-factor(rats$Treatment,levels=1:4,
>labels=c("A","B","C","D"))
```

A5 S-PLUS Functions

S-PLUS has over 3000 in-line functions, a substantial minority of which have been used in this text. Information about these functions is available from the **help** function, for example,

```
>help(paste)
```

Here, we shall illustrate relatively briefly the use of just a small subset of perhaps the most widely used and most useful functions when analysing data. To begin we shall look at the two functions **seq** and **rep**, which are often used in the construction of vectors with a particular pattern of elements. For each function we include a summary of the material given by use of **help**(*function name*)

seq:

Description

Creates a vector of evenly spaced numbers. The beginning, end, spacing, and length of the sequence can be specified.

Usage

seq(...)

seq.default(from=<<see below>>, to=<<see below>>, by=<<see below>>,
   length=<<see below>>, along=NULL)

Optional Arguments

from: Starting value of the sequence. If *to*, *by*, and *length* are all given, the value for this is inferred; otherwise, the default is 1.

to: Ending value of sequence, a value less than from is allowed. If *from*, *by*, and *length* are all given, the value for this is inferred; otherwise, the default is 1.

by: Spacing between successive values in the sequence. If *from*, *to*, and *length* are all given, the value for this is inferred; otherwise, the default is 1.

length: Number of values in the result. If *from*, *to*, and *by* are all given, the value for this is inferred.

along: An object. The length of the object is used as the length of the returned value.

It is an error to specify all of the first four arguments.

Value

A numeric vector with values (from, from+by, from+2*by, ... to); *from* may be larger or smaller than *to*. If *by* is specified, it must have the appropriate sign to generate a finite sequence.

rep

Description

Replicates the input either a certain number of times or to a certain length.

Usage

rep(x, times=<<see below>>, length.out=<<see below>>,
   each=<<see below>>)

Required Arguments

> x vector. Missing values (NAs) are allowed.
> At least one of *times, length.out,* and *each* must be given.

Optional Arguments

times: How many times to replicate x. There are two ways to use *times.* If it is a single value, the whole of x is replicated that many times. If it is a vector that is the same length as x, the result is a vector with times[1] replications of x[1], times[2] of x[2], etc. Zero is allowed in both usages; if times=0, then the length of the result is 0. It is an error if the length of *times* is neither 1 nor the length of x.

length.out: The desired length of the result. This argument may be given instead of *times,* in which case x is replicated as much as needed to produce a result with *length.out* data values. If both *times* and *length.out* are given, *times* is ignored.

each: If *each* is supplied, each element of x is repeated each time. This vector is repeated again if *times* or *length.out* is also given.

Value

A vector of the same mode as x with the data values in x replicated according to the arguments *times, length.out,* and *each.* Any names are removed.

Details

Missing values (NAs) and Infs are treated just like other values.
Some examples of the use of both **seq** and **rep** are

```
> seq(0, 10, length=4)
[1] 0.000000 3.333333 6.666667 10.000000
> seq(0, 50, by=5)
[1] 0 5 10 15 20 25 30 35 40 45 50
> seq(0, 10, by=0.5)
[1] 0.0 0.5 1.0 1.5 2.0 2.5 3.0 3.5 4.0 4.5 5.0 5.5 6.0
[14] 6.5 7.0 7.5 8.0 8.5 9.0 9.5 10.0
```

```
> seq(5, 0, length=3)
[1] 5.0 2.5 0.0
> seq(5)
[1] 1 2 3 4 5
> seq(-5)
[1] 1 0 -1 -2 -3 -4 -5
> rep(10, 5)
[1] 10 10 10 10 10
> rep(c(10, 1), 3)
[1] 10 1 10 1 10 1
> rep(c(10, 1), c(3, 2))
[1] 10 10 10 1 1
> rep(rep(c(10, 1), c(3, 2)), 4)
[1] 10 10 10 1 1 10 10 10 1 1 10 10 10 1 1 10 10 10 1 1
> rep(1:5, 1:5)
[1] 1 2 2 3 3 3 4 4 4 4 5 5 5 5 5
```

Three functions of particular use in statistical work are **mean, var, cor**.

mean

Description

Returns a number that is the mean of the data. A fraction to be trimmed from each end of the ordered data can be specified.

Usage

mean(x, trim=0, na.rm=F)

Required Arguments

x: Numeric object. Missing values (NAs) are allowed.

Optional Arguments

trim: Fraction (between 0 and .5, inclusive) of values to be trimmed from each end of the ordered data. If *trim=.5*, the result is the median.

na.rm: Logical flag: should missing values be removed before computation?

Value

(trimmed) mean of x.

Details

If x contains any NAs, the result will be NA unless na.rm = TRUE.

var and *cor*

Description

Returns the variance of a vector, the variance–covariance (or correlation) matrix of a data matrix, or covariances between matrices or vectors. A trimming fraction may be specified for correlations.

Usage

```
var(x, y, na.method = "fail", unbiased = T, SumSquares = F)
cor(x, y, trim = 0, na.method = "fail", unbiased = T)
```

Required Arguments

x: Numeric matrix or vector, or data frame. May be complex. If a matrix, columns represent variables and rows represent observations. If a data frame, non-numeric variables result in missing values in the result.

Optional Arguments

y: Numeric matrix or vector, or data frame. May be complex. If a matrix, columns represent variables and rows represent observations. If a data frame, non-numeric variables result in missing values in the result. This must have the same number of observations as x.

trim: A number less than .5 giving the proportion trimmed in the internal calculations for cor. This should be a number larger than the suspected fraction of outliers.

na.method: A character string specifying how missing values are to be handled. Options are "fail" (stop if any missing data are found),

"omit" (omit rows with any missing data),

"include" (missing values in the input result in missing values in the output)

"available" (use available observations, see below). Only enough of the string to determine a unique match is required.

unbiased: If TRUE, then variances are sample variances, e.g.,

sum((x-mean(x))^2)/(N-1)

for a vector of length N, which is unbiased if the values in x are obtained by simple random sampling. If FALSE, the definition

sum((x-mean(x))^2)/N

is used instead.

SumSquares: If TRUE, then un-normalized sums of squares are returned, with no division by either N or (N-1) (and unbiased is ignored).

Value

cor() returns correlations, and var() returns variances and covariances or sums of squares.

If x is a matrix, the result is a matrix such that the [i,j] element is the covariance (correlation) of x[,i] and either y[,j] or x[,j]. If x is a vector, the result is a vector, with length equal to the number of columns of y (or length 1 if y is not supplied).

Some examples of the use of these three functions are

```
> x <- 1:10
> mean(x)
[1] 5.5
> var(x)
[1] 9.166667
> x <- c(x, NA)
> mean(x)
[1] NA
> mean(x, na.rm=T)
[1] 5.5
> var(x, na.method="omit")
[1] 9.166667
> X <- matrix(c(1, 4, 6, 3, 2, 1, 4, 5, 6), ncol=3)
```

```
> var(X)
            [,1]    [,2]    [,3]
[1,]    6.333333   -2.5    2.5
[2,]   -2.500000    1.0   -1.0
[3,]    2.500000   -1.0    1.0
> cor(X)
            [,1]            [,2]            [,3]
[1,]   1.0000000    -0.9933993      0.9933993
[2,]  -0.9933993     1.0000000     -1.0000000
[3,]   0.9933993    -1.0000000      1.0000000
```

Two very useful S-PLUS functions for applying a particular S-PLUS function to chosen parts of a matrix or vector are **apply** and **tapply**.

apply

Description

Returns a vector or array by applying a specified function to sections of an array.

Usage

apply(X, MARGIN, FUN, ...)

Required Arguments

X array. Missing values (NAs) are allowed if FUN accepts them.

MARGIN: The subscripts over which the function is to be applied. For example, if X is a matrix, 1 indicates rows, and 2 indicates columns. If the dimensions of X are named, as in the barley data sets, then those names can be used to specify the margin. For a more complex example of the use of MARGIN, see the examples below. Note that MARGIN tells which dimensions of X are retained in the result.

FUN: Function (or character string giving the name of the function) to be applied to the specified array sections. The character form is necessary only for functions with unusual names, e.g., %*%.

Optional Arguments

... any arguments to FUN; they are passed unchanged (including their names) to each call of FUN.

Value

If each call to FUN returns a vector of length N, and N > 1, apply returns an array of dimension

c(N,dim(X)[MARGIN])

If N==1 and MARGIN has length > 1, the value is an array of dimension dim(X)[MARGIN]; otherwise, it is a vector.

tapply

Description

Applies a function to each cell of a ragged array, that is, to the values corresponding to the same levels in all of several categories.

Usage

tapply(X, INDICES, FUN = <<see below>>, ..., simplify = T)

Required Arguments

X vector of data to be grouped by indices. Missing values (NAs) are allowed if FUN accepts them.

INDICES: List whose components are interpreted as categories, each of the same length as X. The elements of the categories define the position in a multiway array corresponding to each X observation. Missing values (NAs) are allowed. The names of INDICES are used as the names of the dimnames of the result. If a vector is given, it will be treated as a list with one component.

Optional Arguments

FUN: Function or character string giving the name of the function to be applied to each cell. If FUN is omitted, **tapply** returns a vector that can be used to subscript the multiway array that **tapply** normally produces. This vector is useful for computing residuals.

Details

Evaluates a function, FUN, on data values that correspond to each cell of a multiway array.

Some examples of the use of these two functions are

```
> apply(X, 2, mean)
[1] NA 1.75 NA
> apply(X, 2, max)
[1] NA 3 NA
> apply(X, 1, min)
[1] NA 2 1 NA
> X <- matrix(c(1, 4, 6, NA, 3, 2, 1, 1, NA, 4, 5, 6,),
ncol=3)
> apply(X, 2, mean)
[1] NA 1.75 NA
> apply(X, 2, mean, na.rm=T)
[1] 3.666667 1.750000 5.000000
> apply(is.na(X), 2, any)#
[1] T F T
> x <- 1:10
> y <- rep(c("NO", "YES"), c(5, 5))
> z <- rep(c("C1", "C2", "C3", "C4", "C5"), 2)
> tapply(x, list(y), max)
NO   YES
5    10
> tapply(x, list(y, z), mean)
      C1  C2  C3  C4  C5
  NO   1   2   3   4   5
 YES   6   7   8   9  10
```

A very useful set of functions in S-PLUS is those relating to probability distributions. For most common probability distributions, S-PLUS has four functions with prefixes d, p, q, and r to return the density (d), cumulative probability (p), quantile (q), or a random sample (r). The following illustrates the use of these functions for the normal distribution.

```
> rnorm(5, 3, 2)
[1] 3.287 3.273 0.132 3.617 2.638
> pnorm(c(-1.96, 0, 1.96))
[1] 0.025 0.500 0.975
> qnorm(c(0.025, 0.5, 0.975))
[1] -1.96 0.00 1.96
> dnorm(c(-1, 0, 1))
[1] 0.242 0.399 0.242
```

A6 Graphics

S-PLUS contains extensive graphical facilities for constructing a variety of plots and diagrams, from the simple to the not-so-simple.

To begin, a device on which to plot has to be specified. Usually this will be a Graphics window. In Windows 98/2000/NT, such a window can be created using the command:

```
>win.graph()
```

We can now create graphics using appropriate S-PLUS commands, with each new graph replacing the previous one. We can also create a postscript file of a plot by using the command:

```
>postscript(file="graph.ps")
```

After producing a postscript file we must close the file using the **dev.off()** command; otherwise, subsequent graphs will be superimposed on the first one.

A large number of graphics parameters control the appearance of a graph. Some of these are listed in Table A4. To find out about additional graphics parameters, use

```
>help(par)
```

Trellis graphics offer access to more-sophisticated diagrams such as coplots and dotplots. A trellis device can be specified for these by use of the command:

```
>trellis.device(win.graph)
```

or

```
>trellis.device(postscript,file = name)
```

Many examples of graphs produced by using the command language are given in the text.

A7 User Functions

S-PLUS commands can be used to write new functions for specific tasks. Although this is a very powerful feature of the software, we give only a few simple examples in this section, merely to illustrate the possibilities.

Table A.4 Commands to Add to Existing Graphs and Plotting Parameters

To add to existing graphs:

Function	Description
points (x, y)	Points at coordinates *x* and *y*
text (x, y, text)	Text at specified coordinates
lines (x, y)	Lines to connect the points given by *x* and *y*
segments (*x*1, *y*1, *x*2, *y*2)	Line segments from (*x*1, *y*1) to (*x*2 to *y*2)
arrows (*x*1, *y*1, *x*2, *y*2)	Arrows segments from (*x*1, *y*1) to (*x*2 to *y*2)
abline (a, b)	Line with intercept *a* and slope *b*
legend (x, y, legend)	Legend
title ("title", "subtitle")	Title at top of figure

Plotting parameters:

Parameter	Purpose
type = "p"/"l"/"h"/"s"/"n", etc	Points/lines/vertical bars/steps/nothing
axes = T/F	With/without axes
main	Main title
sub	Subtitle
xlab, ylab	x/y Axis label
xlim, ylim = c(min, max)	x/y Axis range
pch = 1/2/3, etc. or pch = "+"/".", etc	Plot character
lty = 1/2/3, etc	Line style
lwd	Line width (1 default)

(More complex examples are described at various places in the text.) A detailed account of writing functions in S-PLUS is given in Venables and Ripley (2000).

As a first illustration we shall create a function **mysum**, which calculates the power of the sum of arguments supplied and returns the result:

```
>mysum<-function(a,b,p=1){
 (a+b)^p
{
```

Now **mysum** can be used to take the power of the sum of two numbers:

```
>mysum(1,2)
[1]  3
>mysum(1,2,p=2)
[1]  9
```

The function can also be applied to numerical vectors of the same length:

```
>a<-c(1,2,3,4)
>b<-c(5,6,7,8)
>mysum(a,b,p=3)
[1]  216  512  1000  1728

>mysum(a/2,b/10,p=5)
[1]  1  10  52  172
```

And the function can also be applied to two matrices with the same number of rows and the same number of columns.

```
>a<-matrix(c(1,2,3,4),ncol=2)
>b<-matrix(c(5,6,7,8),ncol=2)
>mysum(a,b,p=3)
        [,1]    [,2]
[1,]    216    1000
[2,]    512    1728
```

The general syntax of a function is

```
function-name<-function (arguments)
{
function body (S-PLUS expressions)
return (output arguments)
}
```

The output arguments are often stored as a list, for example,

```
meanvar<-function(x)
{
m<-mean(x,na.rm=T)
v<-var(x,na.method=omit")
```

```
result<-list(mean=m,variance=v)
}

>meanvar(1:10)
$mean:
[1] 5.5

$variance:
[1] 9.2
```

When writing functions users should try to avoid duplicating the names of existing S-PLUS functions.

Appendix B

Answers to Selected Exercises

Chapter 1

```
#
#
#Exercise 1.1
#
apply(jets[,2:6],2,mean)
#
#Exercise 1.3
#
par(mfrow=c(2,3))
box plot(FFD,ylab="FFD")
box plot(SPR,ylab="SPR")
box plot(RGF,ylab="RGF")
box plot(PLF,ylab="PLF")
box plot(SLF,ylab="SLF")
#
#Exercise 1.4
#
par(mfrow=c(1,1))
```

```
hist(FFD,probability=T)
lines(density(FFD))
```

Chapter 2

```
#
#
#Exercise 2.1
#
plot(husbage,wifeage)
rug(jitter(husbage[!is.na(husbage)]),side=1)
rug(jitter(wifeage[!is.na(wifeage)]),side=2)
#
#
#Exercise 2.5
#
persp(hist2d(husbage,wifeage),xlab="Husband's
age",ylab="Wife's age",zlab="Frequency")
#
#
#Exercise 2.6
#
Access Scatter Plot Matrix dialog from Graphics 2D; click on
Line/Histogram tab and tick draw histogram
#
#
```

Chapter 3

```
#
#
#Exercise 3.1
#
#Here simply use the ANOVA dialog as in the test but use
log(Time)
#as the dependent variable.
#
#
```

```
#Exercise 3.4
#
summary(aov(Response~Status:Condition+Status+Condition))
#
#This gives exactly the same result as
#
summary(aov(Response~Status+Condition+Status:Condition))
#
#S-PLUS does not adjust main effects for the interaction
#composed
#of the main effect factors.
#
#
```

Chapter 5

```
#
#
#Exercise 5.2
#
#
ses<-predict(fit.log,se.fit=T)
#
predict.val<-exp(ses$fit)/(1+exp(ses$fit))
ul<-
exp(ses$fit+1.96*ses$se.fit)/(1+exp(ses$fit+1.96*ses$se.fit)
)
ll<-exp(ses$fit-1.96*ses$se.fit)/(1+exp(ses$fit-1.96*ses$se.fit))
#
ylim<-range(ul,ll)
plot(0:10,predict.val[1:11],xlab="GHQ",ylab="Predicted value
of probability of caseness",
type="l",ylim=ylim)
lines(0:10,ul[1:11],lty=2)
lines(0:10,ll[1:11],lty=2)
legend(locator(1),c("Predicted","95%CI"),lty=1:2)
#
#
```

Chapter 6

```
#
#
#Exercise 6.1
#
t.test(depress.sm[group=="TAU"],depress.sm[group=="BtB"],var
.equal=F)
#
#
#Exercise 6.3
#
#function for filling in missing values in rows of a matrix
by LOCF
#assumes that subject drops out so that a missing value is
not followed
#by any genuine observations
#
locf<-function(X) {
    n<-length(X[,1])
    p<-length(X[1,])
    for(i in 1:n) {
        x<-X[i,]
        y<-x[!is.na(x)]
        N<-length(y)
        if(N!=p) y<-c(y,rep(y[N],p-N))
            X[i,]<-y
    }
    X
}
#
depress.locf<-apply(locf(depress[,4:7]),1,mean)
t.test(depress.locf[group=="TAU"],depress.locf[group=="BtB"]
)
#
#
```

Chapter 7

```
#
#
#Exercise 7.3
#
#
#### function returning - log-likelihood for a mixture of
#two normal distributions
#with both component distributions having the same variance
#
LL<-function(params,data) {
     t1<-dnorm(data,params[2],params[3])
     t2<-dnorm(data,params[4],params[3])
     f<-params[1]*t1+(1-params[1])*t2
     ll<-sum(log(f))
     -ll
}
### fit mixture to geyser data
geyser.res<-
nlminb(c(0.5,50,10,80),LL,data=geyser,lower=c(0.001,-
Inf,0.001,-Inf)
,upper=c(0.999,Inf,Inf,Inf))
#
same<-geyser.res$parameters
#
#assume parameter estimates for different component
variances are in the vector diff
#
x<-seq(40,120,length=100)
#
f1s<-dnorm(x,same[2],same[3])
f2s<-dnorm(x,same[4],same[3])
fs<-same[1]*f1s+(1-same[1])*f2s
#
f1d<-dnorm(x,diff[2],diff[3])
f2d<-dnorm(x,diff[4],diff[5])
fd<-diff[1]*f1d+(1-diff[1])*f2d
```

```
hist(geyser,probability=T,col=0,ylab="Density",ylim=c(0,0.05
),
xlab="Eruption times")
lines(x,fs,lty=1)
lines(x,fd,lty=2)
legend(locator(1),c("Equal variances","Different
variances"),lty=1:2)
#
#
```

Chapter 9

```
#
#
#Exercise 9.3
#
#
skulls.pc<-princomp(skulls[,-1],cor=T)
#
skulls.pc$loadings
#
#
skulls.pcx<-skulls.pc$scores[,1]
skulls.pcy<-skulls.pc$scores[,2]
par(pty="s")
xlim<-range(skulls.pcx)
plot(skulls.pcx,skulls.pcy,xlim=xlim,ylim=xlim,xlab="PC1",yl
ab="PC2",type="n")
labs<-rep(1:5,rep(30,5))
text(skulls.pcx,skulls.pcy,labels=labs)
#
#convex hull
#
for(i in 1:5) {
X<-skulls.pcx[labs==i]
Y<-skulls.pcy[labs==i]
hull<-chull(X,Y)
hull
```

```
polygon(X[hull],Y[hull],density=15,angle=30)
}
#
#
```

Chapter 10

```
#
#
#Exercise 10.1
#
#
n<-length(lowest[,1])
#
#get within group sum of squares regarding data as a single
  cluster
#
wss1<-(n-1)*(var(lowest[,2])+var(lowest[,3])+var(lowest[,4])+
var(lowest[,5]))
#
#apply kmeans for 2 to 6 groups and get within cluster ss
#
wss<-numeric(0)
#
for(i in 2:6) {W<-
sum(kmeans(as.matrix(lowest[,2:5]),i)$withinss)
                 wss<-c(wss,W)
                          }
wss<-c(wss1,wss)
win.graph()
plot(1:6,wss,xlab="Number of clusters",ylab="Within cluster
sum of squares",type="l")
#
#
Exercise 10.3
#
#
#need to look at the eigenvalues in a similar way as for PCA
```

```
#
coords<-cmdscale(congress,k=6,eig=T)
#
plot(1:6,coords$eig,type="l")
#
#
####################################################
######################
```

References

Aitkin, M. (1978). The analysis of unbalanced cross-classifications (with discussion). *Journal of the Royal Statistical Society, A*, 41, 195–223.

Atkinson, A.C. (1987). *Plots, Transformations and Regression*, Oxford Science Publications, Oxford, U.K.

Becker, R.A., Chambers, J.M., and Wilks, A.R. (1988). *New S Language*. Wadsworth and Brook/Cole, Pacific Grove, California.

Burns, K.C. (1984). Motion sickness incidence distribution of time to first emesis and comparison of some complex motion conditions, *Aviation Space and Environmental Medicine*, 56, 521–527.

Chambers, J.M. and Hastie, T.J. (1992). *Statistical Models in S*. Wadsworth and Brook/Cole, Pacific Grove, California.

Chatterjee, S. and Chatterjee, S. (1982). New lamps for old: An exploratory analysis of running times in Olympic Games, *Applied Statistics*, 30, 14–22.

Cleveland, W.S. (1993). *Visualizing Data*, Hobart Press, Summit, New Jersey.

Collett, D. (1991). *Modelling Binary Data*. CRC/Chapman & Hall, London.

Collett, D. (1994). *Modelling Survival Data in Medical Research*. CRC/Chapman & Hall, London.

Cook, R.D. and Weisberg, S. (1982). *Residuals and Influence in Regression*. CRC/Chapman & Hall, London.

Diggle, P.J., Liang, K., and Zeger, S.L. (1994). *Analysis of Longitudinal Data*, Oxford Science Publications, Oxford, U.K.

Efron, B. and Tibshirani, R.J. (1993). *An Introduction to the Bootstrap*. CRC/Chapman & Hall, New York.

Everitt, B.S. (2001). *Statistics for Psychologists*. Lawrence Erlbaum Associates, Mahwah, New Jersey.

Everitt, B.S. (2002). An evaluation of the gap statistic for estimating the number of clusters in a data set. In preparation.

Everitt, B.S. and Dunn, G. (2001). *Applied Multivariate Data Analysis*, 2nd ed. Arnold, London.

Everitt, B.S. and Hand, D.J. (1981). *Finite Mixture Distributions.* CRC/Chapman & Hall, London.

Everitt, B.S. and Pickles, A. (2000). *Statistical Aspects of the Design and Analysis of Clinical Trials.* ICP, London.

Everitt, B.S. and Rabe-Hesketh, S. (1997). *The Analysis of Proximity Data.* Arnold, London.

Everitt, B.S., Landau, S., and Leese, M. (2001). *Cluster Analysis,* 4th ed. Arnold, London.

Goldberg, D. (1972). *The Detection of Psychiatric Illness by Questionnaire.* Oxford University Press, Oxford, U.K.

Hand, D.J. (1998). Discriminant analysis linear. In *Encyclopedia of Biostatistics,* Vol. 2, P. Armitage and T. Colton, Eds., Wiley, Chichester, U.K.

Hand, D.J., Daly, F., Lunn, A.D., McConway, K.J., and Ostrowski, E. (1994). *A Handbook of Small Data Sets.* CRC/Chapman & Hall, London.

Krause, A. and Olson, M. (2000). *The Basics of S and S-PLUS,* 2nd ed. Springer, New York.

McCullagh, P. and Nelder, J.A. (1989). *Generalized Linear Models.* Chapman and Hall, London.

Nelder, J.A., (1977). A reformulation of linear models. *Journal of the Royal Statistical Society, A,* 140, 48–63.

Pinheiro, J.C. and Bates, D.M (2000). *Mixed-Effects Models in S and S-PLUS.* Springer, New York.

Pollock, K.H., Winterstein, S.R., and Conroy, M.J. (1989). Estimation and analysis of survival distributions for radio-tagged animals. *Biometrics,* 45, 99–109.

Rawlings, J.O. (1988). *Applied Regression Analysis.* Wadsworth Books, Pacific Grove, California.

Silverman, B.W. (1986). *Density Estimation for Statistics and Data Analysis.* CRC/Chapman & Hall, London.

Stanley, W. and Miller, M. (1979). Measuring technological change in jet fighter aircraft. Report no. R-2249-AF, Rand Corp., Santa Monica, California.

Therneau, T.M. and Grambach, P.M. (2000). *Modelling Survival Data.* Springer, New York.

Tibshirani, R., Walther, G., and Hastie, T. (2001). Estimating the number of clusters in a data set via the gap statistic. *Journal of the Royal Statistical Society, Series B,* 63, 411–423.

Venables, W.N. and Ripley B.D. (1997). *Modern Applied Statistics with S-PLUS.* Springer, New York.

Index

A

Agglomerative hierarchical methods, 156–158, 165, 169
Analysis of principle components
  for cluster analysis, 158, 163–167
  for multivariate data, 137, 140, 147–151
Analysis of variance (ANOVA)
  analysis of data, 38–50
  description of, 33
  functions of, 70
  longitudinal data, analysis of, 91
  modelling of, 35–38
Arithmetic operators, 7, 37–38, 202–203
Assignments in Command language, 6, 37–38
Assumption of constant variance
  in analysis of variance, 40–41, 43
  in multiple regression analysis, 58–59
Assumption of normality
  in analysis of variance, 40–41, 43
  in logistic regression, 69
  in longitudinal data analysis, 104–106
  in multiple regression analysis, 58–59
Average linkage, 157, 169–171
Axis, labeling of, 74–75, 96–97

B

Balanced data, 35–37, 49
Bandwidth, 175, 188
Baselining, 31, 60, 74–75

Bias corrected accelerated (BCa) confidence
  interval, 113
Binomial distribution, 70, 72, 78
Bivariate density estimation
  analysis of, 176, 186–190
  description of, 173
  modelling of, 173–176
Bootstrap sampling
  for maximum likelihood estimation, 115, 117–120
  for nonlinear regression, 112–113, 114
Box plots
  for analysis of data, 22
  for analysis of variance data, 47, 48
  creation of, 24–25
  example of, 26
  for longitudinal data analysis, 94

C

Censored data, 126–127, 130
Chi-squared
  distribution, 29–30, 72
  model fit, 76
  probability plot, 29–31
  quantiles, 31
Classification rules, 176
Cluster analysis
analysis of data, 158–172
  description of, 155–156
  modelling of, 156–158